Problems in Systematics of Parasites

Proceedings of a Symposium
held under the auspices of the
American Association for the Advancement of Science

Problems in Systematics of Parasites

EDITED BY

GERALD D. SCHMIDT

University Park Press
BALTIMORE, MARYLAND & MANCHESTER, ENGLAND

Library of Congress Catalog Card No. 70–82859
Standard Book No. 8391–0010–6

Printed in the United States of America
By Kingsport Press, Inc.

Contents

Gerald D. Schmidt, Associate Professor of Zoology at Colorado State College, has contributed to the systematics of the Acanthocephala, Nematoda, and Cestoidea, and is currently surveying nematodes of the Pacific islands as potential zoonoses, in association with the Army and Navy.

Gerald D. Schmidt / Introduction

The past few years have seen an upsurge of interest in systematics, particularly in the higher categories. The number of theoretical and practicing systematists is rapidly increasing—sometimes, it seems, in inverse proportion to the number of problems being solved. The fact that young persons are electing careers in the systematics of various groups is a healthy sign, for the problems ahead are formidable. It could be my imagination, but it seems to me that my predecessors classified all the easy species and left only the difficult ones for me to contend with.

In this day of advanced technology many tools have been invented to lighten the burden of the systematist. Never before has he had at his fingertips such an array of techniques and implements designed to ferret out the truth of who is related to whom, and how.

For example, recent advances include high speed data processing by use of computers. It has been suggested that in the near future the laboratory of every systematist will contain a teletype connected to a central computer, probably at the Smithsonian Institution for American systematists, which when fed the appropriate description of a specimen will instantly disgorge its proper identity.

Molecular systematics is now approaching puberty and promises reproducible results for the chemically minded. Based on the axiom that protein structure is the truest reflection of genetic constitution, this approach is yielding striking results in many areas. Researchers in immunology are also contributing to systematics in this regard.

Numerical taxonomy has provided statistical procedures which have been used successfully in the study of several groups, especially arthropods. This controversial discipline has brought about several basic changes in the philosophy of current systematics. Anyone who thinks taxonomy is a dead subject might profitably attend some of the lively discussions that ensue whenever numerical and classical taxonomists get together.

Other comparatively new tools for the systematist include zoogeography, recent and continuing discoveries in paleontology, animal behavior, electron microscopy, and cytogenetics.

And yet, with all of these recent advances at our command there seems to have been no striking revolution, at least in the systematics of parasites. The unique stresses placed on form and function of an organism by a parasitic mode of life have often resulted in unique modifications to which special consideration must be given. The loss of organ systems, for instance, results in fewer observable taxonomic units to be considered. When a species of parasite adapts to different species of hosts, or when it occurs in an atypical host, the adaptive demands encountered often result in structural differences in the organism. Such polymorphism is not found exclusively in parasites, but it certainly remains a pitfall to those who wish to work on their systematics.

When a parasite is removed from its normal environment it ceases to function as a true representative of its species. Yet *in vivo* observations on endoparasites are understandably difficult. *In vitro* cultivation and experimental approaches in general are in their infancy but show much promise in enlightening our ignorance.

A numerical approach to systematics appears fruitless when dealing with polymorphic species or when the sample is restricted to a very few specimens. Genetic studies are difficult in hermaphroditic or asexual organisms.

It is therefore our purpose here to provide a "state of the union" report, in order that we may know where we stand in the systematics of some of the more important parasitic forms. We are fortunate in having contributions from five internationally known authorities, representing five important groups of organisms. What they say here will interest all who are concerned with problems of systematics of parasites.

Wilbur L. Bullock, Professor of Zoology at the University of New Hampshire, has published widely on the histochemistry of parasites and of parasitized and normal tissues, as well as on parasitic protozoa of marine fishes. He is an internationally known authority on the systematics of the Acanthocephala.

Wilbur L. Bullock /

Morphological Features as Tools and as Pitfalls in Acanthocephalan Systematics

> The examination of this small collection of Acanthocephala has led us to attempt a tentative classification of the numerous genera hitherto described. The classification is based largely on the published descriptions of various authors which, unfortunately, are sometimes incomplete in details, a knowledge of which would have been of great assistance, since the simplified morphology of the Acanthocephala offers at best but few characters on which to base a classification, and even these are liable to variation. Our work has therefore been one of great difficulty, and the result leaves much to be desired. To classify the group satisfactorily it will be necessary to obtain a much larger collection of species than we have had at our disposal, and a more extensive knowledge of the life history of the various worms.

With these words Southwell and Macfie began their 1925 paper, "On a collection of Acanthocephala in the Liverpool School of Tropical Medicine." That paper represented the first major attempt to arrange the "Order Acanthocephala"

This study was supported by a research grant (GB-6408) from the National Science Foundation.

into suborders and families. Today, nearly half a century later, we have a larger collection of species available and we have more extensive life history information at our disposal. In addition, we have had the monographs of Meyer (1932–33), Petrotschenko (1956), and Yamaguti (1963). We have also had the numerous publications of Van Cleave from 1914 through 1953 and of Golvan since 1956. Nevertheless, on the basis of lack of agreement among the experts it is clear that the classification of this relatively small group of helminth parasites is as unsatisfactory now as it was in 1925. Comparison of the taxonomic schemes of various workers shows confusion and uncertainty regarding the characteristics of the major groups. The actual names of these groups are not agreed upon nor is there a consensus on whether they are to be considered orders or classes. This confusion extends into the lower taxonomic categories so that species considered by some workers to be in the same genus wind up in different orders or classes of other workers.

There is growing recognition of the importance of ecological, biochemical, and life cycle aspects of helminth parasites for a reliable system of classification, but we are still faced with the same incompleteness and inadequacy of morphological details that discouraged Southwell and Macfie. In spite of a larger collection of species, the simplified, homogeneous morphology of the Acanthocephala leaves us as uncertain in 1968 as our English colleagues were in 1925. The only way we will ever work our way out of this dilemma is to recognize what the problems have been and then attempt to solve them even though the final completion of the task may take several generations. Ecology, biochemistry, and life cycles have much to offer for the future, but we had better examine the ways in which morphology has been used, abused, and misused so that newer methods will really help and not simply add to the confusion. The tool of morphology has at times been a pitfall, and these other methods could become the same traps unless we learn from the experience of the past.

Methods

One of the difficulties of working with Acanthocephala—a difficulty which also occurs in other animal groups, both free living and parasitic—is the problem of technique. Many collections of Acanthocephala have been made as secondary or side projects of other studies. Unfortunately, to produce good whole mounts of Acanthocephala the specimens need special attention from host to microscope slide. It is imperative that the parasites be carefully removed from the host and, prior to fixation, "relaxed" to evert the proboscis completely. Distilled water is the safest and most predictable means to accomplish this relaxation. The time the specimen must spend in distilled water varies from species to species. The best results seem to come when the worms are removed directly from freshly killed hosts. Storage of the worms in Ringer's solution in the refrigerator is satisfactory, but in order to standardize conditions for taxonomic purposes such storage is best avoided. Prolonged standing in distilled water causes blistering of the cuticula and subcuticula.

Hot fixatives often separate the cuticula from the subcuticula; fixation at room temperature is therefore preferred. The choice of fixative does not seem too critical. I routinely use Demke's fixative (alcohol-formalin-acetic acid) but I have had equally good results with Bouin's fluid and neutral buffered 10% formalin.

A wide variety of the usual whole mount stains are satisfactory. I most often use Lynch's precipitated borax carmine technique as outlined in Table 1. I have recently found that chrome alum-gallocyanin often is a fine whole mount stain when nuclear detail is desired. It would seem best, however, to avoid hematoxylin stains, for this substance often causes serious fading.

Besides the use of distilled water to extrude the proboscis, other crucial steps are pricking, dehydration, clearing, and mounting.

Processing Acanthocephala as one would trematodes usually results in shrunken and opaque specimens that are useless. I have found that to guarantee that specimens will not turn chalky white and opaque, it is necessary that each worm be pricked several times with a very fine insect pin such as "minutennadeln." I have made and I have seen specimens that others have made in which this procedure was not used; and some of these specimens were excellent. However, I strongly recommend this tedious process as insurance against the frustration of ending up with useless, opaque worms.

I use a graded series of alcohols to dehydrate specimens. Starting with the weaker solution one or two changes a day are made until absolute alcohol is reached.

If the dehydration or clearing is carried out too rapidly, or if the worms are transferred directly from a nonviscous clearing agent (such as benzene or xylene) into a relatively thick mounting medium, wrinkling of the specimens will occur. Countless hundreds of specimens, including type specimens, are inadequate for taxonomic studies because of this wrinkling. To avoid the shrinkage caused by a viscous mounting medium I use a viscous clearing agent, terpineol. Best results have come from prolonged passage through first 25% terpineol in absolute alcohol, then a 50% solution, then a 75% solution, and finally pure terpineol. The worms can then be transferred directly into a relatively thick resinous mounting medium.

Even with the best of stained whole mounts there are other procedures that need to be considered in order fully to understand acanthocephalan morphology. There is a real need for careful study of serial sections of many species before we can meaningfully discuss the number and morphology of cement glands, the female reproductive tract, the nature of the proboscis receptacle, the structure of the ligament sacs, or the musculature of the presoma or reproductive systems. Linton (1894) condemned the compulsion of many workers to embed and section. He complained: "nature is boiled in corrosive sublimate and fried in paraffin before she is given serious study." However, many of the contradictions and much of the confusion in acanthocephalan morphology could be resolved by a little old fashioned paraffin cookery.

Table 1

Procedure for Processing Acanthocephala for Whole Mounts

1. After careful removal from the host intestine, place worms in distilled water until the proboscis of each is fully extended. This may take from a few minutes to many hours. Difficult specimens can be left in the solution overnight, but only in the refrigerator.
2. Fix in an alcohol-formalin-acetic acid mixture or other fixative for a few hours or overnight.
3. Transfer to 70% ethyl alcohol. (Add 5% glycerin if storage is to be prolonged.)
4. Use entomological "minutennadeln" to prick carefully the body wall of each specimen in several places. Small specimens seem to require this pricking even more than larger ones.
5. In the morning replace alcohol with undiluted Grenacher's borax carmine.
6. In the late afternoon (that is, between 7½ and 8½ hours later) carefully add *one* drop of concentrated hydrochloric acid for each 5 ml of stain. Quickly mix by inverting vial several times. Leave overnight.
7. The next day (preferably in the morning) replace the stain with 1% HCl in 70% alcohol and destain until a light pink color has been obtained. During the destaining process replace the acid alcohol frequently. The destaining process usually requires several hours but may take up to two days. (It is best not to allow the destaining to proceed unattended overnight. Replace acid alcohol with 70% ethyl alcohol and then return to acid alcohol the next day.)
8. After destaining is completed, transfer the specimens to 85% alcohol for 18–24 hours. Be sure to change the alcohol several times to remove excess acid.
9. Dehydrate and clear as follows:
   - 95% alcohol for 6–18 hours
   - 100% alcohol for 6–18 hours
   - 100% alcohol-terpineol sequence
     - 25% terpineol for 6–18 hours
     - 50% terpineol for 6–18 hours
     - 75% terpineol for 6–18 hours
   - Terpineol for 18–24 hours.

   (Note: These terpineol mixtures can be re-used. Decant into a small beaker, allow all debris to settle, and decant into a stock bottle.)
10. Mount worms in thick Permount and apply coverslips.

As emphasized by Cable and Hopp (1954) and West (1964), a careful study of the egg or shelled embryo is best done by using live eggs studied in Ringer's or saline solution. The study of fecal eggs and uterine eggs, as well as of those removed from the body cavity of female worms, helps to insure that mature eggs are being studied.

Finally, it should be emphasized that it is most desirable to base species descriptions on as many specimens as possible. Such series of specimens should ideally include sexually mature males and females as well as immature worms. The establishment of genera and families should, in turn, include a study of the actual specimens as much as possible.

Basic Morphology

The phylum Acanthocephala consists of a homogeneous group of intestinal parasites of relatively simple structure (Fig. 1). These helminths are primarily characterized by the presence of an introvert or proboscis armed with a few to many rows of recurved hooks. The worms completely lack any vestige of organs of digestion. Their life cycles, where known, always involve an arthropod intermediate host.

The acanthocephalan body can be conveniently divided into two general regions: the presoma and the trunk. The presoma is composed of the proboscis, the neck, the proboscis receptacle, the proboscis ganglion, the lemnisci, and the associated muscular structures. The proboscis is that portion of the presoma which is covered with hooks. The neck is that region between the last row of hooks and the cuticular fold; this fold provides a distinct morphological and physiological separation between the subcuticula of the presoma and that of the trunk. The proboscis receptacle is, for the most part, a heavy walled muscular sac into which the proboscis inverts. Situated within this receptacle is the proboscis ganglion or "brain." Arising embryologically from the neck region, and still retaining close association with it in adult individuals, are the lemnisci. These paired structures are peculiar to the Acanthocephala. They are usually surrounded by the neck retractor muscles in their anterior region but hang free in the body cavity for the greater part of their length. A very simple vascular system,

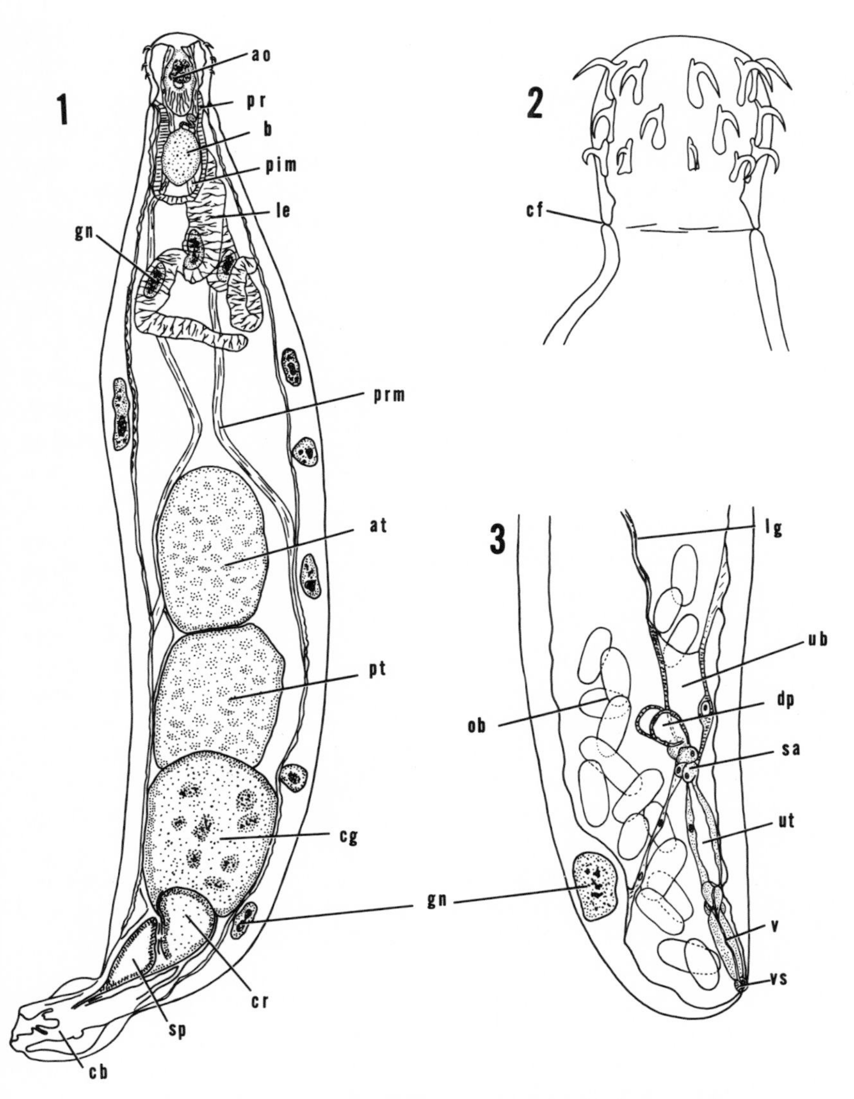

Figure 1. *Octospiniferoides chandleri* Bullock, 1957, from the mosquito fish, *Gambusia affinis.* (1) ENTIRE MALE: *ao*—apical organ; *at*—anterior testis; *b*—proboscis ganglion or "brain"; *cb*—copulatory bursa; *cg*—cement gland; *cr*—cement reservoir; *gn*—giant nucleus; *le*—lemniscus; *pim*—proboscis invertor muscle; *prm*—proboscis retractor muscle; *pr*—proboscis receptacle; *pt*—posterior testis; *sp*—seminal vesicle. (2) PROBOSCIS AND NECK: *cf*—cuticular fold. (3) POSTERIOR END OF FEMALE: *dp*—dorsal pouch; *lg*—ligament; *ob*—ovarian balls; *sa*—selective apparatus; *ub*—uterine bell; *ut*—uterus; *v*—vagina; *vs*—vaginal sphincter.

the lacunar system, is embedded within the subcuticula of the trunk. This system appears to be without definite walls. It is composed of one or two main longitudinal vessels and a number of connecting transverse vessels that may result in some degree of pseudosegmentation or form an irregular network.

The female gonad breaks up early in development into numerous fragments, the ovarian balls. Within these ovarian balls the mature ova are fertilized, and after detaching themselves undergo many cleavages in the body of the female. Upon reaching maturity the embryos pass through the genital ducts to the outside. The anterior portion of these ducts is a funnel shaped structure, the uterine bell. At the posterior, narrow base of the uterine bell are a group of cells, the selective apparatus, and usually a pair of dorsal pouches. The selective apparatus and one or more openings into the pseudocoel appear to regulate which eggs are passed on into the tubular uterus and which are returned to the body cavity. Although Yamaguti (1963) has challenged this interpretation on the basis of finding immature eggs in the uterus, it still seems to be the best explanation for the function of these cells. Although this explanation admittedly needs more substantial proof, the fact that the structure does not function perfectly does not preclude this explanation. From the uterus the eggs are passed through the vagina, a tubular structure having several specialized cells associated with it. Some of these cells appear to be glandular; others appear to be muscular and to function as sphincters.

The reproductive organs of the male are more complicated. A pair of testes is usually arranged in tandem. Just behind the testes are the cement glands. These vary in number from a single syncytial gland in *Octospiniferoides* and other Eoacanthocephala to eight glands in the Archiacanthocephala and some Palaeacanthocephala. They vary in shape from spherical to very elongately tubular. The sperm ducts from the testes have several enlargements for the storage of sperm. Both the sperm ducts and the cement gland ducts enter the penis. An extrusible copulatory bursa surrounds the penis. An elongate muscular sac, Saefftigen's pouch, lies parallel to the reproductive ducts and probably functions in the extrusion of the bursa.

The body cavity of the trunk is divided longitudinally by membranes that enclose the ligament sacs. These sacs may be single or double; if double they occur as a dorsal sac and a ventral sac. In some acanthocephalans they may persist throughout life; in others they may break up during the sexual development of the worms.

The acanthocephalan life cycle is typically a two host life cycle. When ingested by the arthropod intermediate host (insect or crustacean) the acanthor hatches and penetrates the gut wall. Within the hemocoel of the arthropod the acanthella develops and eventually transforms into a cystacanth or juvenile. In some life cycles paratenic hosts are present and in some cases these may be essential enough to be considered second intermediate hosts. Although numerous life cycles have been completed in recent years there are still many genera and even families for which no life cycles are known. Here is an area for much fruitful research.

While the above mentioned organs, along with others, are of considerable importance for separating species and genera, only a few of these structures have been used to define the higher taxa. Among characteristics that are frequently used to describe orders and classes are: the location of the longitudinal vessels of the lacunar system, the nature of the ligament sacs, the nature and number of the cement glands, the structure of the proboscis receptacle, the spination of the acanthor, the nature of the embryonic membranes, and the presence or absence of trunk spines. These features will now be considered in some detail in an attempt to evaluate their usefulness as characteristics of higher taxonomic units within the phylum.

Lacunar System

Meyer (1932) considered the position of the main longitudinal vessels of the lacunar system to be of prime importance as a distinguishing characteristic between the two orders he had established in 1931. In the Archiacanthocephala there was usually a dorsal and ventral vessel or only a dorsal; in the Palaeacanthocephala there was usually a pair of lateral vessels. Most of the exceptions in Meyer's scheme were cared for with the establishment of the Eoacanthocephala by Van Cleave (1936). Since that time the Archiacanthocephala and the Eoacanthocephala have been characterized by the dorsal-ventral location and the Palaeacanthocephala by the lateral position.

The details of the lacunar system are unfortunately not always easy to see. Consequently many descriptions of species, genera, and even families do not note the location of the main longitudinal vessels or other details of this system. Thus this aspect of the Meyer-Van Cleave taxonomic system has been generally accepted on the one hand or ignored on the other; it has not been carefully substantiated nor has it been discredited. Golvan (1959, 1960, 1962) incorporates this feature into his descriptions of the classes and some of the families of the phylum. Yamaguti (1963) characterizes the lacunar system of two of his orders (Apororhynchidea and Echinorhynchidea) and ignores the system in the description of the other two (Neoechinorhynchidea and Gigantorhynchidea). He does, however, spottily refer to the nature of the system in families and genera, presumably on the basis of available information.

In the Palaeacanthocephala the main longitudinal vessels are usually located laterally, but in some forms the whole system becomes so irregular and network-like that even in sections it is difficult to identify the main vessels with certainty. One apparently difficult genus is *Polyacanthorhynchus*. Baylis (1927), in his description of *P. macrorhynchus* (Diesing, 1851), refers to the lacunar system as "branching." Meyer (1932) does not mention it. Golvan (1962), however, indicates that it is dorsal and ventral and not lateral but he gives no firm evidence for his description. On the basis of this and other characters Golvan found this to be the only genus he could not assign to one of the three orders of the Meyer-Van Cleave system.

In the Eoacanthocephala the longitudinal vessels are usually distinctly dorsal and ventral. The position of these vessels is further marked by the giant nuclei which are intimately associated with them. This is certainly the case in genera such as *Neoechinorhynchus, Octospinifer,* and *Octospiniferoides* that I have studied in serial section. However, in some genera, such as *Quadrigyrus, Pallisentis,* and *Pandosentis,* the main vessels and nuclei are usually located laterally near the posterior end of the worm. In the Archiacanthocephala a survey of the species descriptions given by Meyer shows that the lacunar system has been described in less than half of the species and genera. Among the forms having both a dorsal and a ventral longitudinal vessel are species of *Apororhynchus, Oligacanthorhynchus, Nephridiacanthus, Travassosia, Macracanthorhynchus, Pachysentis, Oncicola,* and *Echinopardalis.* Forms having only a dorsal vessel are species of *Mediorhynchus* and *Gigantorhynchus.* In *Moniliformis* Meyer indicates that there is only a dorsal vessel in *M. moniliformis* but both a dorsal and a ventral vessel in *M. gracilis* and *M. travassosi.* In *Prosthenorchis, P. elegans* has only a dorsal vessel, *P. luhei* has both, and *P. sigmoides* sometimes has only one and sometimes has both!

Golvan (1962) describes the principal vessels of the lacunar system as "*dorsal et ventral (parfois un seul canal dorsal)*." However, he divides his class Archiacanthocephala into two orders, and both orders are described as having dorsal *and* ventral vessels. I have not found any indication as to which genera have "*un seul canal dorsal*"—even in Golvan's descriptions of the genera so indicated by Meyer. Obviously, more work needs to be done on the histology of Acanthocephala before the systematic usefulness of this system can be established. However, enough is known to see considerable validity to the criteria of Meyer and Van Cleave as at least a suggestive pattern.

Ligament and Ligament Sacs

The most complete descriptions of the ligament and ligament sacs have appeared in studies of archiacanthocephalans of the family Oligacanthorhynchidae. In *Macracanthorhynchus* and in *Oligacanthorhynchus,* for example, there are two ligament sacs, a dorsal and a ventral, both of which are persistent throughout the life of the worm. Anteriorly these sacs end on the posterior end of the proboscis receptacle; posteriorly they end around the reproductive organs. In the female the dorsal sac surrounds the anterior openings of the uterine bell and the ventral sac connects with the female genital ducts through the posterior ventral openings of the selective apparatus. All workers agree that the Archiacanthocephala have persistent ligament sacs with the dorsal sac terminating posteriorly at the uterine bell. However, since the nature of these sacs can be accurately determined only by a study of serial sections, there are many genera and species in which this feature has not been studied.

There is similar agreement that the ligament sac in the Palaeacanthocephala is single and that this sac ruptures early in the sexual development of these worms. The anterior attachment in this group is at the posterior end of the proboscis receptacle and the posterior attachment is on the reproductive organs. In the female this posterior attachment is inside the base of the uterine bell. Since this single ligament sac encloses the reproductive organs in immature worms it is usually observed clearly in well stained whole mounts. Its subsequent disintegration is likewise readily observed. However, even here there are numerous genera and families in which this structure has never been studied.

In the Eoacanthocephala there appear to be both dorsal and ventral sacs as in the Archiacanthocephala. However, these sacs do not persist in those members of this group that have been studied (for example, *Neoechinorhynchus*, *Octospiniferoides*, and *Tanaorhamphus* as studied by myself). Furthermore, in the Eoacanthocephala the uterine bell is associated with the ventral ligament sac.

The information we have on the comparative anatomy of the ligament sacs is still too fragmentary to provide conclusive evidence of ordinal or class affiliation. However, there seems to be a pattern that might be a significant aid to understanding the morphological characteristics of the higher taxa.

Subcuticular Nuclei

As pointed out by Van Cleave (1928) and confirmed by many workers since, the nuclei of the subcuticula can be traced through a pattern of development that is also reflected in the systematics of the phylum. The early acanthella of all species studied thus far have simple, vesicular nuclei associated with the developing integument. In the Eoacanthocephala these nuclei remain, without further division, in usually precise small numbers in equally precise locations. In *Gracilisentis* the nuclei remain spherical, in *Neoechinorhynchus* they are irregular in outline ("ameboid"), whereas in *Quadrigyrus* they become branched. In the Archiacanthocephala the nuclei become elongate and branched (*Macracanthorhynchus*) or they may fragment with the fragments remaining in close association with each other (*Moniliformis* and *Mediorhynchus*). In a few of the Palaeacanthocephala, such as *Leptorhynchoides,* the nuclei are much branched with some fragmentation, but in most of this group there is extensive fragmentation of the nuclei with the fragments becoming widely scattered throughout the subcuticula. In at least two genera, *Arhythmorhynchus* and *Centrorhynchus,* these fragments are restricted to a portion of the anterior trunk so that the subcuticula of much of the trunk is nearly devoid of nuclear material.

Obviously there is a pattern in the nature and distribution of subcuticular nuclei which can be correlated with the other characteristics in ordinal classification. However, the pattern is not rigid. With the exception of the Eoacanthocephala, the nature of the subcuticular nuclei, although of considerable importance at the family and generic level, cannot be used as a single decisive character for determining ordinal assignment.

Proboscis Receptacle

There appears to be a relatively stable pattern in the structure of the wall of the proboscis receptacle. In the Palaeacanthocephala the receptacle wall is composed of two distinct muscle layers so that the receptacle is a sac within a sac. In the Archiacanthocephala there is only one thick muscle layer and this layer usually has a prominent ventral cleft the greater part of its length. In the Eoacanthocephala the receptacle is a simple, one layered, closed sac. In all cases there are accessory muscles that surround the receptacle but the homologues have never been ascertained.

A few problems exist, however, in using the structure of the proboscis receptacle as a single diagnostic character for making class or ordinal assignments. Yamaguti (1963) and others have considered the Apororhynchidae, with the single genus *Apororhynchus,* to represent a distinct order on the basis of the absence of any proboscis receptacle. However, as has been pointed out by Van Cleave (1952), the basic structure of this genus is that of the Archiacanthocephala. The cement glands are the typical eight, uninucleate glands; the lacunar vessels conform to the archiacanthocephalan pattern, as do the ligament sacs and the subcuticular nuclei. Therefore, it seems best to go along with Van Cleave and Petrotschenko, reject a separate ordinal assignment, and consider the Apororhynchidae as highly specialized archiacanthocephalans.

The family Moniliformidae, although it resembles the rest of the Archiacanthocephala in most aspects of its basic morphology, has a receptacle that is double walled and which lacks the ventral cleft that occurs in other members of the order. Furthermore, the outer wall of the receptacle consists of a prominent spiral muscle layer that does not seem to have been reported for any other acanthocephalan. As in the case of the Apororhynchidae, there does not seem to be any valid reason for assigning the Moniliformidae to any other order. Rather, in characterizing the Archiacanthocephala these two departures from the typical body organization need to be recorded.

Descriptions of species in the genus *Mediorhynchus* have indicated that the receptacle wall of some species is single while others have a double wall in the anterior portion. On the strength of these varying descriptions Yamaguti (1963) has set up the genus *Empodisma* for those species that have a double anterior wall to the receptacle. It is my opinion, after studying the written descriptions of other authors, their illustrations, and specimens of two of the species from North America (*M. robustus* and *M. papillosus*) that the receptacle is always basically single walled but that there is a close association of the extrinsic presomal muscles with the anterior portion of the receptacle. Obviously such an opinion needs to be substantiated by detailed morphological studies of serial sectioned material from a number of forms. Meanwhile it seems ill advised to retain the genus *Empodisma* on such tenuous morphological grounds.

Petrotschenko (1956) erected and Yamaguti (1963) accepted the subfamily Polyacanthorhynchinae of the Rhadinorhynchidae. The distinguishing feature of this subfamily was the presence of a single walled receptacle in members of a family and order in which this structure was typically double walled. The genus *Polyacanthorhynchus* has been one of the knottiest problems of ordinal assignment in the phylum. Baylis (1927), after a careful study of *P. macrorhynchus* from *Arapaima gigas* concluded, largely on the basis of trunk spines and a single walled receptacle, that *Polyacanthorhynchus* was related to *Quadrigyrus*. However, Baylis did not properly interpret either the multinucleate nature of the cement gland or the nature of the subcuticular giant nuclei as described by Van Cleave (1920) for *Quadrigyrus*. In the light of present knowledge it is inconceivable that anyone could suggest that these two species might be congeneric. Much more recently, Golvan (1962), after listing the various morphological features attributed to *Polyacanthorhynchus*, stated: "*Nous considérons donc, provisoirement, l'assignation du genre* Polyacanthorhynchus *à l'une des 3 Classes du Phylum comme incertaine. Une étude anatomique complémentaire, faite sur un matériel récolté dans de bonnes conditiones s'impose.*" Study of some of the original Baylis material leads me to agree with Petrotschenko and Yamaguti that this genus is basically a rhadinorhynchid. Pending further study, therefore, it seems best to accept this genus as recommended by Petrotschenko and Yamaguti.

Petrotschenko (1956) set up the genus *Protorhadinorhynchus* to contain *P. ditrematis* (Yamaguti, 1939) and *P. carangis* (Yamaguti, 1939). Both of these species were originally assigned to *Rhadinorhynchus* by Yamaguti, but he indicated in the descriptions that both species had receptacles with only a single muscle wall. There was no mention as to how this unique feature was determined and actually his Figure 32 of *R. ditrematis* shows a typical double walled receptacle of a typical rhadinorhynchid. It would seem that further study is necessary before either of these species can be given definite generic assignment. Until this is done the status of *Protorhadinorhynchus* is in doubt.

In general, it seems that the proboscis receptacle is a reliable character for characterizing three major groups of Acanthocephala. The most important exceptions are *Moniliformis, Polyacanthorhynchus,* and *Apororhynchus.* Other presumed exceptions might merely be misinterpretations on the basis of lack of detailed morphological studies.

Cement Glands

The shape, number, and structure of the cement glands are morphological features of the Acanthocephala that have been used by Lühe (1911), Meyer (1932–1933), Van Cleave (1936, 1948, and 1949), and Golvan (1959, 1960, and 1962). These authors have used the cement glands as characteristics for most of the taxonomic units within the phylum. Unfortunately, in some cases they tended to be quite rigid. As a result of this rigidity—together with stronger emphasis on other morphological features—several workers have de-emphasized these glands. Thus Southwell and Macfie (1925) considered the number and shape of cement glands as too variable. On similar grounds Thapar (1927) concluded that use of the cement glands resulted in an artificial system. Petrotschenko (1956) and Yamaguti (1963) assign a minor role to the cement gland as either an ordinal or familial characteristic.

Use of the cement glands to develop a natural system of classification of the Acanthocephala involves several practical difficulties. A considerable number of species and genera reported in the literature have not yet been characterized on the basis of these glands. In some cases this omission is due to the male of the species being yet unknown. In other cases we are handicapped by the fact that earlier workers did not pay much attention to such details; most of their descriptions completely lack any reference to internal anatomy.

Most confusing of all are those instances where the number of cement glands has been erroneously reported due to misinterpretation of the number of glands or because of unjustified inferences about the condition of the glands. Misinterpretation is easy in some genera in which the cement glands are long, tubular, and intertwined. Likewise if nontubular cement glands are bunched together it is often difficult to determine accurately the number of glands. Often cross sections of the worms at the appropriate level are decisive, but occasionally even this will not readily solve the problem. The unjustified inferences are the most difficult to recognize and there are numerous instances of completely unnatural groupings as a result. For example, Van Cleave and Lincicome (1940) separated the family Gorgorhynchidae with four cement glands from the Rhadinorhynchidae with eight glands. Cable and Linderoth (1963) have shown that the type species of the type genus of the latter family (*Rhadinorhynchus pristis*) actually has four tubular cement glands. Thus the distinction between these families is destroyed at its most critical point. Chandler (1934) erected the genus *Nipporhynchus* on the basis of its having four cement glands. He separated it from *Rhadinorhynchus* by assuming that "if the type *pristis* had other than the usual eight [cement glands], the fact would have been mentioned by Lühe." Actually Lühe made no reference to the number of such glands.

Another example of this type of unjustified inference occurs likewise among the rhadinorhynchid acanthocephalans. When Van Cleave (1923) set up the genus *Telosentis* he made no reference to the cement glands as he was most impressed by the genital spines. This preoccupation with genital spines led him (1947) to include Linton's *Rhadinorhynchus tenuicornis,* a form with eight cement glands, in the genus *Telosentis*. Apparently Van Cleave had earlier made this association since he and Lincicome (1940) list *Telosentis* as a member of the Rhadinorhynchidae because it possesses eight cement glands. I have recently been studying the type material for *Telosentis molini,* the genotype, and have determined that the males have only *four* cement glands. Therefore we have further evidence for deciding that, as previously suggested by Cable and Linderoth (1963), *Telosentis tenuicornis* cannot possibly be congeneric with *T. molini.*

This leads us to the main question as to whether, in spite of all of the confusion due to mistakes, omissions, and misinterpretations, the cement glands are of any value in acanthocephalan taxonomy, and if so, at what levels. To investigate this problem it is basic to consider Van Cleave's (1949) discussion of the morphological and phylogenetic interpretation of the cement glands (Fig. 2). Petrotschenko

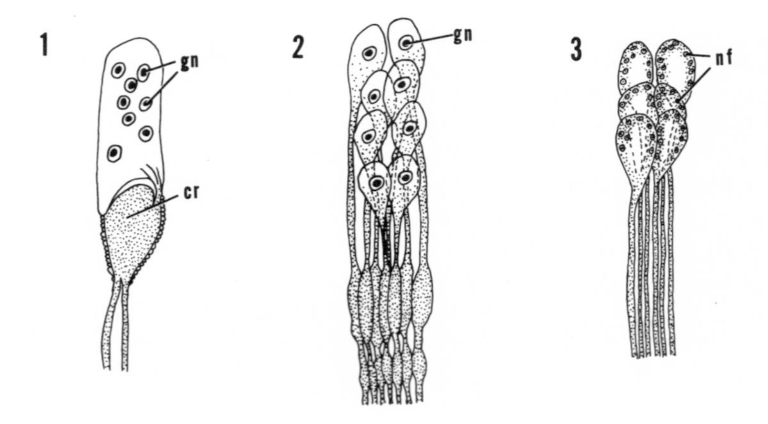

Figure 2. Acanthocephalan cement glands. (1) EOACANTHOCEPHALA: single syncytial cement gland with giant nuclei (*gn*) and a separate cement reservoir (*cr*). (2) ARCHIACANTHOCEPHALA: usually eight cement glands, each with a single giant nucleus (*gn*). (3) PALAEACANTHOCEPHALA: several glands, each of which is syncytial with numerous nuclear fragments. (Six pear-shaped glands are shown but the number may vary from two to eight and the shape may vary from spherical to elongately tubular.)
(*Entire figure redrawn from Van Cleave, 1949*)

and Yamaguti appear to have disregarded this important study of the basic morphology of acanthocephalan cement glands. Van Cleave notes that there are three basic morphological types of cement glands within the phylum. There is the single syncytial cement gland of the Eoacanthocephala. This organ type has several giant nuclei in the gland, and the gland is followed by a nonnucleate cement reservoir. The single syncytial cement gland is recognized by Yamaguti as one of the features of his Neoechinorhynchidea. The only possible exception is the genus *Acanthogyrus* for which two syncytial cement glands have been described by both the original author (Thapar, 1927) and subsequent workers. Although Dollfus and Golvan (1956) suggest that Thapar confused the cement reservoir with a second cement gland, it seems best to accept this one minor exception to the otherwise homogeneous picture of cement gland structure in this branch of the phylum.

Van Cleave also showed that the cement glands of the remaining members of the phylum could be further arranged into two distinct subdivisions. In those acanthocephalans of terrestrial hosts (Archiacanthocephala) the cement glands are characteristically unicellular. The nuclei are of the ameboid, giant type with a large, prominent nucleolus. In acanthocephalans with this morphology there are nearly always eight glands. Only in the genera *Mediorhynchus* and *Moniliformis* have species been described in which there are other than the usual eight. Van Cleave (1949) concluded that such departures are the result of inaccurate counts (in both genera the glands are closely crowded together) or of the occurrence of teratological individuals. Perhaps there may be some intraspecific variability in these genera. I know of no careful study of this problem since 1949 that furnishes any substantial evidence for disagreeing with Van Cleave in this conclusion.

Some workers have emphasized the diversity of cement gland counts in the Palaeacanthocephala. And certainly more diversity occurs in this branch of the phylum than is found in the other groups. Unfortunately, emphasis on the variability in the number, as well as the shape, of the cement glands loses sight of the characteristic palaeacanthocephalan morphology of the glands of this group. In contrast to the uninucleate cement glands of the Archiacanthocephala, the glands of the Palaeacanthocephala are always multinucleate whether they are round or tubular, whether there are only four, or whether there are eight.

It seems to me that these morphological features substantiate the basic Meyer-Van Cleave system so long as we recognize that the number of cement glands is of class or ordinal value in the Eoacanthocephala and probably the Archiacanthocephala. In the Palaeacanthocephala these features are merely family, perhaps generic, or even only specific values. On the other hand, the cytological features, when clearly seen, are diagnostic for the three groups: Eoacanthocephala, Archiacanthocephala, and Palaeacanthocephala. Such a recognition of the basic, nonadaptive aspects of cement gland morphology avoids the artificial assignment of *Centrorhynchus* and *Prosthorhynchus* to a relationship with the Archiacanthocephala on the basis of less reliable, adaptive characters such as the nature of the egg membrane. Proper understanding of these basic cytological types clearly associates *Centrorhynchus* with the polymorphid Palaeacanthocephala and associates *Mediorhynchus* with the gigantorhynchid Archiacanthocephala. Such a separation is supported by other morphological features which require acceptance of the unique insertion of the proboscis receptacle (in the middle of the proboscis) as a case of convergence.

Acanthor Membranes

Many of the references to the membranes surrounding the acanthors indicate the presence of three membranes. However, West (1964) has given clear evidence of the presence of four membranes, a view postulated earlier by Manter (1928) and Meyer (1928) (Fig. 3). The outer membrane is a thin covering that is actually the membrane of the unfertilized ovum. Beneath this membrane is the fibrillar coat (Monné and Hönig, 1954). Underneath the fibrillar coat is the thickened fertilization membrane. The innermost membrane is thin and closely applied to the surface of the acanthor. Much of the confusion has occurred from assuming that the "shell" of a land egg such as in *Macracanthorhynchus hirudinaceus* is the same as the "shell" of an aquatic form such as *Echinorhynchus gadi*. As West clearly points out, the prominent feature of the aquatic egg is the much thickened fertilization membrane, a structure that frequently is drawn out at the poles in the form of "polar elongations." The "shell" of the land egg is the fibrillar coat, a membrane that is often much reduced in aquatic eggs (for example, *Acanthocephalus jacksoni*). In addition the outermost coat is often rubbed off in the fecal eggs of land forms. Petrotschenko (1956) and Yamaguti (1963) have placed

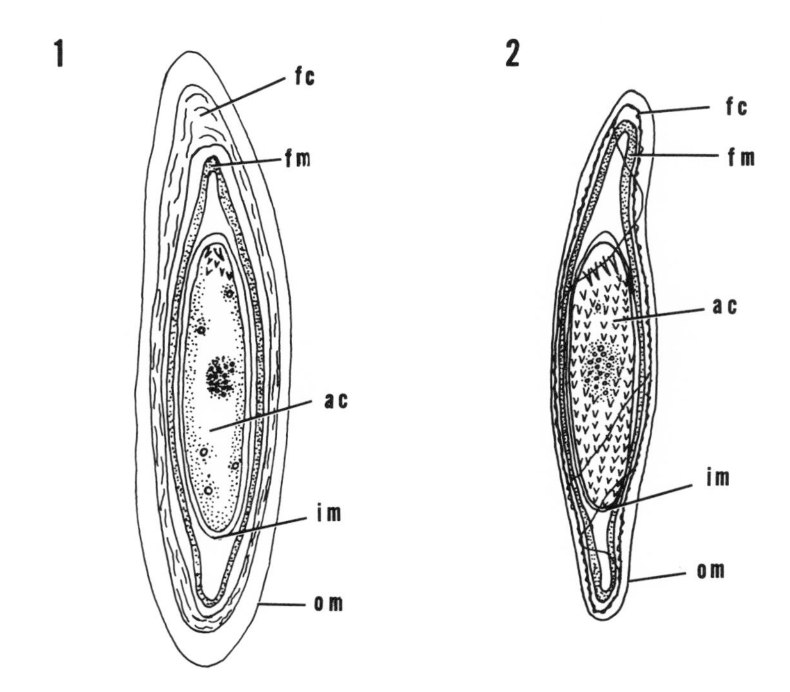

Figure 3. Acanthor membranes. (1) *Echinorhynchus gadi* (Zoega) Mueller, 1776. (2) *Acanthocephalus jacksoni* Bullock, 1962.
ac—acanthor; *fc*—fibrillar coat; *fm*—fertilization membrane; *im*—inner membrane; *om*—outer membrane.

great emphasis on the morphological features of the acanthocephalan egg as an ordinal characteristic. Therefore all forms with land eggs are included in the Gigantorhynchidea and excluded from the aquatic Echinorhynchidea even though they are morphologically similar in other respects. Such an emphasis on these egg characteristics leads to a drastic realignment of genera and families. Thus, on the basis of their having a land egg Petrotschenko erected the family Pseudoacanthocephalidae for the new genus *Pseudoacanthocephalus,* the genus to contain *P. bufonis* and two related species from toads. This family was then placed in the Order Gigantorhynchidea whereas all of the other species of *Acanthocephalus* were retained in the Echinorhynchidea. But the only differences between these two groups are the nature of the egg membranes and the habitat of the amphibian hosts. Similarly, these same two workers place the genus *Prosthorhynchus,* along with *Porrorchis* and *Pseudoporrorchis,* in the Prosthorhynchidae of Petrotschenko (1956), and this family in turn is included in the Gigantorhynchidea. The genus *Plagiorhynchus* is retained in the Plagiorhynchidae, a family in the Echinorhynchidea. No consideration is given either to life cycle patterns, which admittedly are poorly known, or to other morphological features. Similar emphasis on the egg membranes caused these authors to include the Centrorhynchidae in the Gigantorhynchidea.

The nature of the egg membrane is an adaptive character, and is therefore not a valid morphological feature for characterizing families and orders unless it can be supported by more substantial evidence. This principle is well expressed by Mayr, Linsley, and Usinger (1953) :

> Linnaeus and most of his followers for nearly a century classified birds by purely adaptational characters. Birds with webbed feet were put into one category; birds with a hooked bill were considered another group; etc. Eventually it was realized that characters that are adaptations to a specific mode of living are not only subject to rapid changes by selective forces, but may also be acquired in different unrelated lines. Such characters have only limited value in establish-

> ing taxonomic categories. They are mostly useful in separating species and genera. When dealing with the classification of higher categories we must search for characters that tend to remain stable, characters that are phylogenetically conservative.

The *Plagiorhynchus-Prosthorhynchus* problem is particularly enlightening. These two genera have been confused for years in terms of their true morphological distinctives. Van Cleave (1942) emphasized the source of confusion on the basis of the location of the brain; the brain was considered as being in the posterior extremity of the proboscis receptacle in *Prosthorhynchus* and in the middle third in *Plagiorhynchus*. This character has proved to be an exceedingly difficult one since it is almost impossible to see the brain in stained specimens. Golvan (1956) looked at more than a hundred specimens of various species of both genera after the specimens were cleared, without staining, in chloral-lacto-phenol. He found that the brain was always located in the middle third of the receptacle. Therefore this character was no longer valid. However, Golvan considered that the genera could still be separated on the basis of the habitat of the host and the type of egg associated with this habitat. Then, as indicated above, on the basis of extreme reliance on these same characters Petrotschenko and Yamaguti in their monographs placed each genus in a different family and a different order. Recently Schmidt and Kuntz (1966), after pointing out that the host habitat is not an invariable character and that the type of egg is not strictly a habitat feature, concluded that the differences between these two groups of species is only at the subgeneric level. They therefore reduced the genus *Prosthorhynchus* to subgeneric status within the genus *Plagiorhynchus*. Such a move, if justified, obviously eliminates the Prosthorhynchidae as a separate family and does away with the incongruity of placing two groups of obviously related species in separate orders. Such taxonomic lumping is further supported by the crustacean nature of the intermediate host for *P. formosus* (by Sinitzin, 1929, and by Schmidt and Olsen, 1964) and for *P. cylindraceus* (by Dollfus and Dalens, 1960). The development of these land related acanthocephalans in aquatic related intermediate hosts is strong evidence of their palaeacanthocephalan affinities. From this it seems reasonable to conclude that the nature of the egg membranes cannot be used as an ordinal or even family or generic characteristic unless there is substantial morphological and/or life cycle evidence to support it.

Acanthor Morphology

The larval form that develops within the egg membranes of the Acanthocephala is the acanthor. It is this larva that hatches from the egg, upon ingestion by the appropriate intermediate host, and which then develops within the hemocoel. The acanthor typically has a compact mass of darkly staining, small nuclei in the central region, called the embryonalkern, and a few large, vesicular nuclei in the cortical region. There are numerous reports of spines on the acanthor, as well as descriptions in which spines are not mentioned. There are a few reports of the absence of spines. When these structures have been described they usually involve a small number of large spines (hooks) anteriorly; small spines may be limited to a short region just behind the field of hooks or they may extend to the posterior end of the acanthor.

Both Petrotschenko (1956) and Yamaguti (1963) make much of the nature of the spination of the acanthor as a character of ordinal importance. Thus the keys of both these authors refer to the "unarmed embryo" of the Neoechinorhynchidea; to the hooks on one end only of the embryo of the Echinorhynchidea; and to the hooks and spines that completely cover the embryo of the Gigantorhynchidea. Several investigators in recent years have demonstrated that such strong reliance on this characteristic is unjustified. Golvan and Deltour (1964) demonstrated that the embryos of *Pseudoporrorchis centropi, Corynosoma villosus, Acanthocephalus madagascarensis*, and *Arhythmorhynchus* sp. are all armed with hooks anteriorly and with spines over the remainder of the acanthor. Of these species, only *Pseudoporrorchis* is inclined in the Gigantorhynchidea of Petrotschenko and Yamaguti. Obviously, either the other three genera are misplaced or the character is not valid. Grabda-Kazubska (1964) described with most precise detail the armature of the acanthors of 13 species of Palaeacanthocephala, including *Acanthocephalus*, *Echinorhynchus*, *Pomphorhynchus*, *Polymorphus*, *Filicollis*, and *Centrorhynchus*. She also redescribed the armature for the acanthor of *Macracanthorhynchus hirudinaceus*. All of these species had both anterior hooks and posterior spines; the latter covered the entire acanthor. In *Echinorhynchus*, and especially in *E. gadi*, these spines were very small and only "weakly visible." This may account for the lack of reference to them in previous studies of this common species, including the careful observations of West (1964). In the other species the spines were readily visible. From these observations it seems clear that the acanthor of the Palaeacanthocephala (Echinorhynchidea) cannot be characterized as having only anterior hooks and lacking body spines.

The Eoacanthocephala present other problems. Meyer (1931*b*) failed to notice any hooks or spines on the acanthor of *Neoechinorhynchus rutili*. His observations were confirmed by Grabda-Kazubska (1964) although the latter admitted that she may not have been able to study fully developed acanthors. Ward (1940) made no mention of acanthor spines in her description of the life cycle stages of *Neoechinorhynchus cylindratus*. However, acanthor spines have been described and figured for several other eoacanthocephalans in the course of studies on their life cycles (Fig. 4). Thus spines have been shown to occur on the acanthors of *Neoechinorhynchus emydis* (by Hopp, 1954), *Octospinifer macilentis* (by Harms, 1965), and *Paulisentis fractus* (by Cable and Dill, 1967). Unpublished observations of my own on *Octospiniferoides chandleri* and with one of my students on *Tanaorhamphus ambiguus* provide additional support for the presence of spines on a significant number of neoechinorhynchid acanthors. The study of Grabda-Kazubska (1964) has indicated that acanthor spination may be an important taxonomic aid at the species and genus level. Such studies should therefore be continued. However, acanthor spines do not appear to be of systematic value for the major taxa of Acanthocephala.

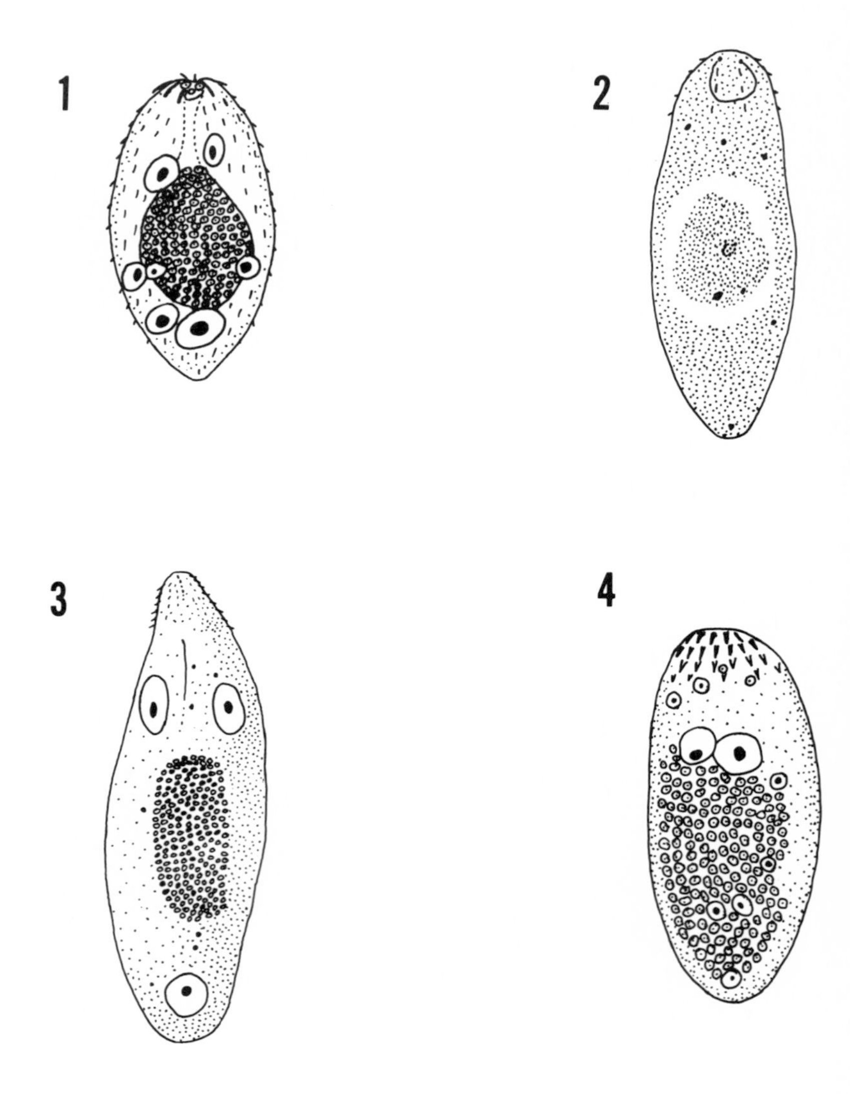

Figure 4. Acanthors of four species of Eoacanthocephala showing presence of spines (1) *Paulisentis fractus* (*After Cable and Dill, 1967*) (2) *Neoechinorhynchus emydis* (*After Hopp, 1954*) (3) *Octospinifer macilentis* (*After Harms, 1965*) (4) *Tanaorhamphus ambiguus*.

Trunk Spines

All species of Archiacanthocephala completely lack trunk spines. In both the Eoacanthocephala and the Palaeacanthocephala there are forms with and forms without such spines. In the Eoacanthocephala the species with trunk spines are all in the *Acanthogyrus-Pallisentis-Quadrigyrus* group, a group which Van Cleave (1948) incorporated into the order Gyracanthocephala in the class Eoacanthocephala. The order Neoacanthocephala was established for the forms without spines, mostly members of the family Neoechinorhynchidae. These two orders, systematically separated as they are by only the single character of the presence or absence of spines, have not been generally accepted by many workers. Although it seems to be significant that the Gyracanthocephala have undergone most of their speciation in Asia, while the Neoacanthocephala have undergone theirs in North America, it is probably best to restrict this combination of geographical distribution with the presence or absence of spines to the family level for the present. Unfortunately, many of the species and genera involved in this problem are only poorly known. There seems to be, however, no question but that the presence or absence of spines is a good family character.

In the Palaeacanthocephala the picture is much more complicated. Trunk spines seem to have been developed in several independent lines. Thus there seems to be some evidence for a basic polymorphid morphological pattern with trunk spines in *Polymorphus, Corynosoma, Arhythmorhynchus* but no spines in *Centrorhynchus* or *Plagiorhynchus*. Is the presence or absence of spines important enough for us to erect separate families? Perhaps, but certainly the feature is not one of ordinal or even suborder importance, as could be eventually the case in the Eoacanthocephala. The situation is still more confusing in the echinorhynchid-rhadinorhynchid group which contains numerous genera and families. But the use of trunk spines, even when attempts are made to combine this character with the number and shape of cement glands, has not stabilized any valid system of classification for these worms. So we find

Golvan and Houin (1964) considering *Fessisentis, Filisoma* (both without spines), and *Gorgorhynchus* (with spines) to be in the same family on the basis that each has four cement glands. Yamaguti, on the other hand, included *Leptorhynchoides* (usually considered to be a rhadinorhynchid without spines) in the Echinorhynchidae because it lacked trunk spines. It does not, then, seem that the trunk spines can be more than a supporting feature for recognition of families in the Palaeacanthocephala. Certainly at the ordinal level the best we can say is: "spines present or absent."

Conclusions

Prior to Hamann (1892) all species of acanthocephalans were commonly assigned to the genus *Echinorhynchus*. With the recognition of the genera *Gigantorhynchus* and *Neorhynchus* (= *Neoechinorhynchus*), along with establishment of three families (Echinorhynchidae, Gigantorhynchidae, and Neorhynchidae), Hamann initiated the modern era of acanthocephalan systematics. During the succeeding 30 years, as distinctive patterns of acanthocephalan morphology were recognized, not only new species but also numerous genera and families were described. The first attempt to propose a classification arranging families into suborders (of the single order Acanthocephala) was that of Southwell and Macfie (1925). These authors essentially raised the families of Hamann to suborder status. However, their misunderstanding of some basic features of acanthocephalan morphology led to some artificial and untenable associations. Thus they included *Apororhynchus* in the Neoechinorhynchidea, *Mediorhynchus* and *Empodius* (along with *Centrorhynchus*) in the Echinorhynchidea, and *Moniliformis* in the Echinorhynchidea. They recognized the importance of the proboscis receptacle, the cement glands, and the subcuticular nuclei, but they failed to apply these characters accurately in their scheme of families and suborders.

Although he made no attempt to establish orders or suborders, Travassos (1926) arranged the known genera into family groupings that foreshadowed the Meyer-Van Cleave system. He pointed out, for example, the incongruity of grouping *Mediorhynchus* and *Centrorhynchus* in the same family, as had been done by Van Cleave (1916) and by Southwell and Macfie. Thapar (1927), on the other hand, established three orders on the basis of the presence of trunk spines (Acanthogyridea), the absence of trunk spines (Echinorhynchidea), and the absence of both trunk spines and proboscis hooks (Apororhynchidea). This emphasis on a superficial feature (trunk spines) led to an exceedingly artificial system.

Meyer (1931*a*) divided the class Acanthocephala into two orders, Palaeacanthocephala and Archiacanthocephala, on the basis of: (1) the lacunar system, (2) the arrangement of proboscis hooks, (3) the presence of protonephridial organs in a few forms, (4) the nature of the ligament sacs, and (5) the habitat of the host. Van Cleave (1936) removed the inconsistencies in Meyer's system by the establishment of a third order, the Eoacanthocephala, on the basis of: (1) giant nuclei in the subcuticula, (2) single syncytial cement gland, and (3) a single-walled proboscis receptacle. Golvan (1959, 1960, 1962) basically followed this system.

The orders Neoechinorhynchidea, Echinorhynchidea, and Gigantorhynchidea of Petrotschenko and Yamaguti roughly correspond to the Eoacanthocephala, Palaeacanthocephala, and Archiacanthocephala respectively of the Meyer-Van Cleave system. However, as I indicated earlier there are numerous instances in which genera and families are unrealistically assigned to orders on the basis of overemphasis on adaptive characteristics. For this reason the Meyer-Van Cleave system is the better of the commonly used systems, even though it must be modified to bring it up to date and to make it more consistent. There are numerous incompatible and irreconcilable ways in which the names Neoechinorhynchidea, Echinorhynchidea, and Gigantorhynchidea have been used by Southwell and Macfie (1925), by Thapar (1927), by Petrotschenko (1956), by Yamaguti (1963), and by others. It would appear, therefore, that acanthocephalan systematics would best be served by a suppression of these names, shrouded in ambiguity as they are, in favor of the much clearer understanding of the Acanthocephala involved in the names and characterization of the Eoacanthocephala, the Palaeacanthocephala, and the Archiacanthocephala.

It therefore seems best to characterize three orders of Acanthocephala in a system similar to that resulting from the establishment of the order Eoacanthocephala by Van Cleave (1936). Table 2 indicates the manner in which the Meyer-Van Cleave system can be brought up to date in the light of the problems I discussed earlier.

The subdivision of these orders into suborders and families is just as tentative now as it was in the days of Southwell and Macfie, Thapar, and Travassos. On the basis of known information there does not appear to be any merit in grouping Palaeacanthocephala and Archiacanthocephala together as Van Cleave (1949) did with his class Metacanthocephala. Neither does there seem to be much justification for dividing the Archiacanthocephala into any suborders such as suggested by Golvan (1962). It seems best to encompass the known genera into four families: Oligacanthorhynchidae, Moniliformidae, Giganthorhynchidae, and Apororhynchidae.

As indicated earlier, Van Cleave (1949) divided his class Eoacanthocephala into the orders Gyracanthocephala and Neoacanthocephala on the basis of the presence of trunk spines in the former group. There may be some merit in reverting to Van Cleave's (1936) concept of these groups as suborders, but much more work needs to be done, especially in the Gyracanthocephala, before the unity of this group can be proved or disproved. Certainly the status of families such as Acanthogyridae, Quadrigyridae, and Pallisentidae needs further study. We need detailed morphological studies and we must elucidate some life cycles in order to determine the relationships of these families to each other and to the other Eoacanthocephala. In the neoacanthocephalan group most of the genera seem naturally to form the family Neoechinorhynchidae. The genera *Tenuisentis, Pandosentis,* and *Tanaorhamphus* have been associated into a separate family (Tenuisentidae), but there seems to be little justification for this at this time. Neither does there seem to be any real reason for recognition of a separate family for the genus

Table 2

Characterization of Orders in the Phylum Acanthocephala

Character	Palaeacanthocephala	Archiacanthocephala	Eoacanthocephala
Body size	From small to large	Mostly large	Small
Host habitat	Mostly aquatic	Terrestrial	Aquatic
Lacunar system—main longitudinal vessels	Generally lateral	Dorsal and ventral or dorsal only	Dorsal and ventral, at least anteriorly
Cement glands	From two to eight, multinucleate	Usually (always?) eight, uninucleate	Usually one, syncytial, with giant nuclei; distinct cement reservoir
Trunk spines	Present or absent	Absent	Present or absent
Subcuticular nuclei	Numerous amitotic fragments or few highly branched	Few, elongate or branched, or with fragments remaining close together	Very few giant nuclei
Proboscis receptacle	Closed sac with two muscle layers, except in Polyacanthorhynchinae	Single muscle layer, often modified by ventral cleft or accessory muscles	Closed sac with single muscle layer
Ligament sacs	Single. Ruptured in mature worms. Posterior attachment inside uterine bell	Dorsal and ventral. Persistent. Dorsal sac attaches to uterine bell	Dorsal and ventral. Disappear in adult. Ventral sac attaches to uterine bell
Nephridia	Absent	Present or absent	Absent
Embryonic membranes	Usually thin	Usually thick	Thin
Intermediate hosts	Crustacea	Insects (and millipedes)	Crustacea

Hebesoma; indeed, further study may show that *H. violentum* is only another species of *Neoechinorhynchus*. Golvan (1959), Petrotschenko (1956), and Yamaguti (1963) have attempted to divide the family Neoechinorhynchidae into several subfamilies. As we lack morphological details and life cycle information for most genera such subdividing seems premature.

The systematics of the Palaeacanthocephala is confused, and sometimes approaches a state of chaos. From a survey of the various proposals in the literature as well as the morphological characteristics of the numerous genera there seem to be three major groupings of genera: the polymorphids, the echinorhynchids, and the rhadinorhynchids. The polymorphids include such genera as *Polymorphus, Corynosoma,* and *Arhythmorhynchus*. More peripherally associated with this group are *Centrorhynchus* and the Plagiorhynchidae; this latter group shows morphological features approaching the echinorhynchids. In the echinorhynchids we have *Echinorhynchus, Acanthocephalus,* and *Pomphorhynchus* as typical genera with similar morphology. In the rhadinorhynchid group there are *Rhadinorhynchus, Gorgorhynchus, Telosentis, Tegorhynchus,* and *Serrasentis*. Other genera, such as *Fessisentis, Filisoma,* and *Leptorhynchoides,* show intermediate conditions of various types between typical echinorhynchids and typical rhadinorhynchids. As described in this chapter, the confusing array of morphological features, especially cement gland shape and number and the presence or absence of trunk spines, makes any definitive classification of the Palaeacanthocephala impossible at this time. More information is needed on life cycles and embryology. More biochemical data would help. But more information on basic morphology must be assembled. All of these studies must be performed with care and in detail. Then, after we have a clearer conception of the separation of adaptive from nonadaptive features, we should be able to use morphology and other methods as genuine taxonomic aids.

REFERENCES

Baylis, H. A. 1927. Some parasitic worms from *Arapaima gigas* (Teleostean fish) with a description of *Philometra senticosa* n. sp. (Filaroidea). *Parasitology* 19: 35–47.

Bullock, W. L. 1966. A redescription of *Octospiniferoides chandleri* Bullock, 1957. *J. Parasitol.* 52: 735–738.

Cable, R. M., and Dill, W. T. 1967. The morphology and life history of *Paulisentis fractus* Van Cleave and Bangham, 1949 (Acanthocephala: Neoechinorhynchidae). *J. Parasitol.* 53: 810–817.

——— and Hopp, W. B. 1954. Acanthocephalan parasites of the genus *Neoechinorhynchus* in North American turtles with the descriptions of two new species. *J. Parasitol.* 40: 674–680.

——— and Linderoth, J. 1963. Taxonomy of some Acanthocephala from marine fishes with reference to species from Curacao, N. A. and Jamaica, W. I. *J. Parasitol.* 49: 706–716.

Chandler, A. C. 1934. A revision of the genus *Rhadinorhynchus* (Acanthocephala) with descriptions of new genera and species. *Parasitology* 26: 351–358.

Dollfus, R. P., and Dalens, H. 1960. *Prosthorhynchus cylindraceus* (Goeze, 1782) au stade juvénile chez un isopode terrestre. Acanthocephala-Polymorphidae. *Ann. Parasitol. Hum. Comp.* 35: 347–349.

——— and Golvan, Y. 1956. Acanthocéphales de Poissons du Niger. *Bull. Inst. Franc. D'Afr. Noire* 18: 1086–1109.

Golvan, Y. J. 1956. Acanthocéphales d'oiseaux. Troisième note. Revision des espèces Européennes de la sous-famille de Plagiorhynchinae A. Meyer 1931 (Polymorphidae). *Ann. Parasitol. Hum. Comp.* 31: 350–384.

——— 1959. Le phylum des Acanthocephala. Deuxième note. La classe des Eoacanthocephala (Van Cleave, 1936). *Ann. Parasitol. Hum. Comp.* 34: 5–52.

——— 1960–1961. Le phylum des Acanthocephala. Troisième note. La classe des Palaeacanthocephala (Meyer, 1931). *Ann. Parasitol. Hum. Comp.* 35: 138–165, 350–386, 573–593, 713–723; 36: 76–91, 612–647, 717–736.

——— 1962. Le phylum Acanthocephala (Quatrième note) La classe des Archiacanthocephala (A. Meyer, 1931). *Ann. Parasitol. Hum. Comp.* 37: 1–72.

——— and Deltour, F. 1964. Spinulation des larves acanthors d'Acanthocéphales: consequences phylogeniques et systematiques. *C. R. Acad. Sci. Paris* 258: 4355–4357.

——— and Houin, R. 1964. Revision des Palaeacanthocephala. Deuxième note. Famille des Gorgorhynchidae Van Cleave and Lincicome, 1940. *Ann. Parasitol. Hum. Comp.* 39: 535–605.

Grabda-Kazubska, B. 1964. Observations on the armature of embryos of acanthocephalans. *Acta Parasitol. Polonica* 12: 215–231.

Hamann, O. 1892. Das System der Acanthocephalen. *Zool. Anz.* 15: 195–197.

Harms, C. E. 1965. The life cycle and larval development of *Octospinifer macilentus* (Acanthocephala: Neoechinorhynchidae). *J. Parasitol.* 51: 286–293.

Hopp, W. B. 1954. Studies on the morphology and life cycle of *Neoechinorhynchus emydis* (Leidy), an acanthocephalan parasite of the map turtle, *Graptemys geographica* (Le Sueur). *J. Parasitol.* 40: 284–299.

Linton, E. 1894. Some observations concerning fish-parasites. *Bull. U. S. Fish Comm. for 1893.* pp. 101–112.

Lühe, M. 1911. *Die Süsswasserfauna Deutschlands.* Vol. XVI. *Acanthocephalen.* Verlag Fischer, Jena.

Manter, H. W. 1928. Notes on the eggs and larvae of the thorny headed worm of hogs. *Trans. Amer. Micr. Soc.* 47: 342–347.

Mayr, E., Linsley, E. G., and Usinger, R. L. 1953. *Methods and Principles of Systematic Zoology.* McGraw-Hill, New York.

Meyer, A. 1928. Die Furchung nebst Eibildung, Reifung und Befruchtung des *Gigantorhynchus gigas.* Ein Beitrag zur Morphologie der Acanthocephalen. *Zool. Jahrb., Abt. Anat.* 50: 117–218.

——— 1931*a*. Neue Acanthocephalen aus dem Berliner Museum. *Zool. Jahrb., Abt. Syst.* 62: 53–108.

——— 1931*b*. Urhautzelle, Hautbahn und plasmodiale Entwicklung der Larve von *Neoechinorhynchus rutili* (Acanthocephala). *Zool. Jahrb., Abt. Anat.* 53: 103–126.

——— 1932–1933. Acanthocephala. *Bronn's Klassen und Ordnungen des Tierreichs.* Band IV, Abt. 2, Buch 2. Akademische Verlagsgesellschaft M.B.H., Leipzig.

Monné, L., and Hönig, G. 1954. On the embryonic envelopes of *Polymorphus botulus* and *P. minutus* (Acanthocephala). *Ark. Zool.* Ser. 2, 7: 257–260.

Petrotschenko, V. I. 1956. *Acanthocephala of Wild and Domestic Animals.* Vol. I. Akademiya Nauk S.S.S.R., Moscow. 435 pp. [in Russian].

Schmidt, G. D., and Kuntz, R. E. 1966. New and little-known plagiorhynchid Acanthocephala from Taiwan and the Pescadores Islands. *J. Parasitol.* 52: 520–527.

——— and Olsen, O. W. 1964. Life cycle and development of *Prosthorhynchus formosus* (Van Cleave, 1918) Travassos, 1926, an acanthocephalan parasite of birds. *J. Parasitol.* 50: 721–730.

Sinitzin, D. 1929. Note on an intermediate host for *Plagiorhynchus formosus. J. Parasitol.* 15: 287.

Southwell, T., and Macfie, J. W. S. 1925. On a collection of Acanthocephala in the Liverpool School of Tropical Medicine. *Ann. Trop. Med. Parasitol.* 19: 141–184.

Thapar, G. S. 1927. On *Acanthogyrus* n.g. from the intestine of the Indian fish *Labeo rohita,* with a note on the classification of the Acanthocephala. *J. Helminthol.* 5: 109–120.

Travassos, L. 1926. Contribuções para o conhecimento da fauna helminthologica brasileira. XX. Revisão dos Acanthocephalos brasileiros. Parte II. Familia Echinorhynchidae Hamann, 1892, sub-fam. Centrorhynchinae Travassos, 1919. *Mem. Inst. Oswaldo Cruz* 19: 31–125.

Van Cleave, H. J. 1916. Acanthocephala of the genera *Centrorhynchus* and *Mediorhynchus* (new genus) from North American birds. *Trans. Amer. Microscop. Soc.* 35: 221–232.

——— 1920. Two new genera and species of acanthocephalous worms from Venezuelan fishes. *Proc. U. S. Natl. Mus.* 58: 455–466.

——— 1923. *Telosentis,* a new genus of Acanthocephala from southern Europe. *J. Parasitol.* 9: 174–175.

——— 1928. Nuclei of the subcuticula in the Acanthocephala. *Zeitschr. Zellforsch. Mikroscop. Anat.* 7: 109–113.

——— 1936. The recognition of a new order in the Acanthocephala. *J. Parasitol.* 22: 202–206.

——— 1942. A reconsideration of *Plagiorhynchus formosus* and observations on Acanthocephala with atypical lemnisci. *Trans. Amer. Microscop. Soc.* 61: 206–210.

——— 1948. Expanding horizons in the recognition of a phylum. *J. Parasitol.* 34: 1–20.

——— 1949. Morphological and phylogenetic interpretations of the cement glands in the Acanthocephala. *J. Morphol.* 84: 427–457.

——— 1952. Speciation and formation of genera in Acanthocephala. *Syst. Zool.* 1: 72–83.

——— and Lincicome, D. R. 1940. A reconsideration of the acanthocephalan family Rhadinorhynchidae. *J. Parasitol.* 26: 75–81.

Ward, H. L. 1940. Studies on the life history of *Neoechinorhynchus cylindratus. Trans. Amer. Microscop. Soc.* 59: 327–347.

West, A. J. 1964. The acanthor membranes of two species of Acanthocephala. *J. Parasitol.* 50: 731–734.

Yamaguti, S. 1939. Studies on the helminth fauna of Japan. Part 29. Acanthocephala II. *Jap. J. Zool.* 8: 317–351.

——— 1963. *Systema Helminthum.* Vol. V. *Acanthocephala.* Interscience, New York.

Discussion

Question: Dr. Bullock, I do not think you mentioned the name of the class to which you would assign the three orders Eoacanthocephala, Palaeacanthocephala, and Archiacanthocephala.

Dr. Bullock: This is because I have come to no decision on this point.

Question: You have done considerable work on the biochemistry of Acanthocephala, but has there been any use for this in your systematics?

Dr. Bullock: Actually I have not done anything in biochemistry. Years ago I did some work in histochemistry and Dr. Schmidt's reference to my contributions in that field deals with ancient history, mainly because I became involved with straightening out some of the morphology and taxonomy of the Acanthocephala before using any of these newer techniques. I think such techniques may be useful but we have so many problems due to inadequate descriptions of species that these first need to be checked by conventional methods, including, in many instances, serial sectioning.

Question: Did you say, Dr. Bullock, that you have seen posterior spines on the acanthor?

Dr. Bullock: Yes.

Question: This is probably far-fetched, but did you ever consider the possibility of an evolutionary relationship between the Monogenea with their posterior haptor with hooks and spines and the Acanthocephala?

Dr. Bullock: No, I guess I have been so busy working on the morphology and taxonomy of the Acanthocephala that I have not spent much time worrying about where they fit in the animal kingdom.

Question: Do you feel that the cement glands of the land forms are specific in number for a given species?

Dr. Bullock: If by land forms you mean the systematic group Archiacanthocephala, it seems to be specific. As you know, there are deviations in number reported in the literature. Van Cleave's 1945 paper suggested that these were either errors in counting because they were crowded or that they represented teratological individuals.

Question: Of course, when an author says there were errors in counting that is an easy way out. I have seen specimens and in fact have some in my collection which I am certain have variations in numbers of cement glands. I have dissected some and I think there are variations within the species. I think descriptions should mention the variations, such as "from six to eight glands," or whatever.

Dr. Bullock: Well, certainly on the basis of my work with the genus *Acanthocephalus* I heartily agree. I still think that the typical number of cement glands for *Acanthocephalus* is six, even though in *Acanthocephalus jacksoni,* which I worked on a few years ago, we found anywhere from four to eleven. I cannot remember what the average was now, but the vast majority of them had six. I have not done much direct work with the Archiacanthocephala to see what kind of variation there is within that group, so as far as characterizing the cement glands within the Archiacanthocephala is concerned I am willing to go along with a qualifying "usually."

Question: Yes, this would be better, for I think there are these exceptions where, for some reason or other, specimens from the same host show variations.

Dr. Bullock: This is still a little different from saying that the presence of six glands is characteristic of one species, while eight is characteristic of another species, which definitely is the case within the Palaeacanthocephala, even when there is intraspecific variation. When I get some Archiacanthocephala I usually pass them along to Gerald Schmidt.

Marietta Voge, Associate Professor of Medical Microbiology at the University of California School of Medicine at Los Angeles, is widely known for her work on the ecology, distribution, and systematics of cestodes and has contributed many advances to in vitro *cultivation of larval forms.*

Marietta Voge / Systematics of Cestodes—Present and Future

> Get your facts first, and then you can distort them as much as you please.
>
> Mark Twain (Samuel Clemens)

I remember my great pleasure when I described my first new species. The description still stands, probably because taxonomists are fewer, or less critical. Now that I have been forced to take a good look at the manner in which cestode organisms are classified, I am disappointed and somewhat annoyed, because to my mind no significant progress or change has occurred in cestode systematics in the last 15 years. Most importantly, perhaps, there has been no attempt to evaluate the system critically or to question it in terms of present day knowledge.

As is true for other parasitic metazoa, the basis for cestode taxonomy is morphology and, to a lesser extent, host distribution. In some groups life history information is added to morphologic description of the adult stages. Assignment to ordinal and familial categories rests chiefly on the structure of the attachment organs on the scolex. As in other helminths, descriptions of genera and species frequently are based on relatively few specimens from a single host individual, although there has been an improvement in this respect in recent years. The extent of morphologic variability has been studied in very few species only.

Supported in part by Research Grant USPHS AI-07332, from the National Institutes of Health, U. S. Public Health Service.

In spite of the compilation of extensive data and the publication of several volumes on cestode taxonomy (Spassky, 1951; Yamaguti, 1959), no significant procedural changes have taken place, even though a substantial amount of new knowledge has accumulated. Proposals for some rearrangements by different workers, even though based on reasonable evidence, have so far not been accepted by taxonomists of the older generation. A few examples will be discussed below.

Recent studies on the fundamental biology of cestodes, with bearing on systematics, include the utilization of carbohydrates by different cestode species (see Read and Simmons, 1963), the effect of cestodes on their hosts (Mueller, 1965; 1968), the structure and composition of calcareous corpuscles (von Brand *et al.*, 1967), temperature tolerances of apparently closely related species, and life histories with detailed descriptions of developmental sequences and larval stages (various authors).

Our concept of host specificity, until recently rather vague and lacking in explanatory facts, has received more precise backing from investigations on the internal environment of hosts and adaptations of their parasites (Read *et al.*, 1959; Simmons, 1961; Read, 1968). The capabilities and limitations of certain cestodes, as observed *in vitro*, have provided additional characteristics useful for specific differentiation. Studies on the excystment of larval stages (Rothman, 1959) and on the hatching of oncospheres (Berntzen and Voge, 1965), demonstrated the differential action of host enzymes and other substances upon different cestode species. Experiments with host intestinal emptying time (Read and Voge, 1954) demonstrated one of the factors which can prevent establishment of a cestode in an otherwise satisfactory definitive host.

In the intermediate host, establishment of the oncosphere is dependent upon hatching under appropriate conditions in the host intestine. The structure of the intestinal wall must allow for penetration of the hatched oncosphere and its passage to the host body cavity (Voge and Graiwer, 1964). Although most of this information is still fragmentary, continued studies on the interaction between host and parasite will provide us with a sounder basis for judging the importance of "host specificity" as a taxonomic criterion, and will incidentally yield important information about the parasite itself.

In the following pages I will briefly discuss the major past and present criteria used for the grouping of cestode species into orders, families, and genera, and will give examples of apparent inconsistencies or deficiencies in our procedures. I am particularly concerned with the meagerness of data used to back up the present system, even though sufficient information is available to support redistribution of at least some cestode groups.

I will also speculate a little on the future of cestode systematics and on the possibility of using characters other than morphological in the definition and grouping of species.

Ordinal Classification

The number of cestode orders varies somewhat with the opinions of different taxonomists. However, most students agree on the six major orders which contain the majority of the known species. The structure of the scolex, the prime basis for classification, is shown in Figure 1. The orders Tetrarhynchidea and Tetraphyllidea, in their adult phase, are restricted to elasmobranchs; the Pseudophyllidea occur in marine and freshwater teleosts, and also in mammals and birds; the Caryophyllidea are found in freshwater fish and annelids; and the Proteocephalidea are seen primarily in freshwater fish, amphibia, and reptiles. With few exceptions the Cyclophyllidea occur in terrestrial vertebrates, particularly mammals and birds.

At the ordinal level the greatest structural homogeneity is observed in the groups restricted to marine and freshwater hosts. Thus, the characters of the scolex of tetrarhynchids, tetraphyllids, and caryophyllids have a high "predictive value" for host distribution and such additional features as are known for these groups. Structural uniformity of the scolex is also present in pseudophyllids and proteocephalids, although many species in the latter group could be assigned to the Cyclophyllidea if the scolex were the only criterion used. Structural diversity of the adult stages is generally low and assignment to subordinal taxa is therefore a difficult matter.

The greatest structural diversity of the scolex, and of the adult strobila, is seen in the groups presently encompassed by the Cyclophyllidea.

Although the scolex in all species of the order bears four suckers, it also may be variously modified by the presence of hooks or a rostellum. The number, structure, and distribution of the hooks may vary widely. Moreover, as has been mentioned before, the Proteocephalidea also are characterized by a scolex with four suckers. Host distribution in cyclophyllids is variable and by itself does not serve to differentiate this order from all the others. The only consistent internal character, the compactness of the vitellaria or yolk glands, is hardly of sufficient importance to offset the variability of other attributes. If one of our aims is consistency, it should be noted that the vitellaria are follicular in several of the other orders so that, in those, the vitelline structure is not useful in ordinal separation.

If practicality is one of the important criteria of taxonomic procedure, the cyclophyllid assemblage leaves much to be desired. Even though appropriate subdivision at the family level alleviates the inconsistencies in the characterization of higher taxa, many of the cyclophyllid families encompass within themselves several heterogeneous groups. As will be discussed below, some of these assemblages are not practical, nor do they make biological sense. Among the relatively homogeneous families are the Mesocestoididae, Nematotaeniidae, Hymenolepididae, and Taeniidae. Their overall structural and biological characteristics are, by comparison with presently used ordinal characteristics, of sufficient magnitude to warrant elevation to ordinal rank should this be desirable. Some of the remaining families, particularly the Anoplocephalidae and Dilepididae, present considerable problems by their heterogeneity and consequent lack of useful definition.

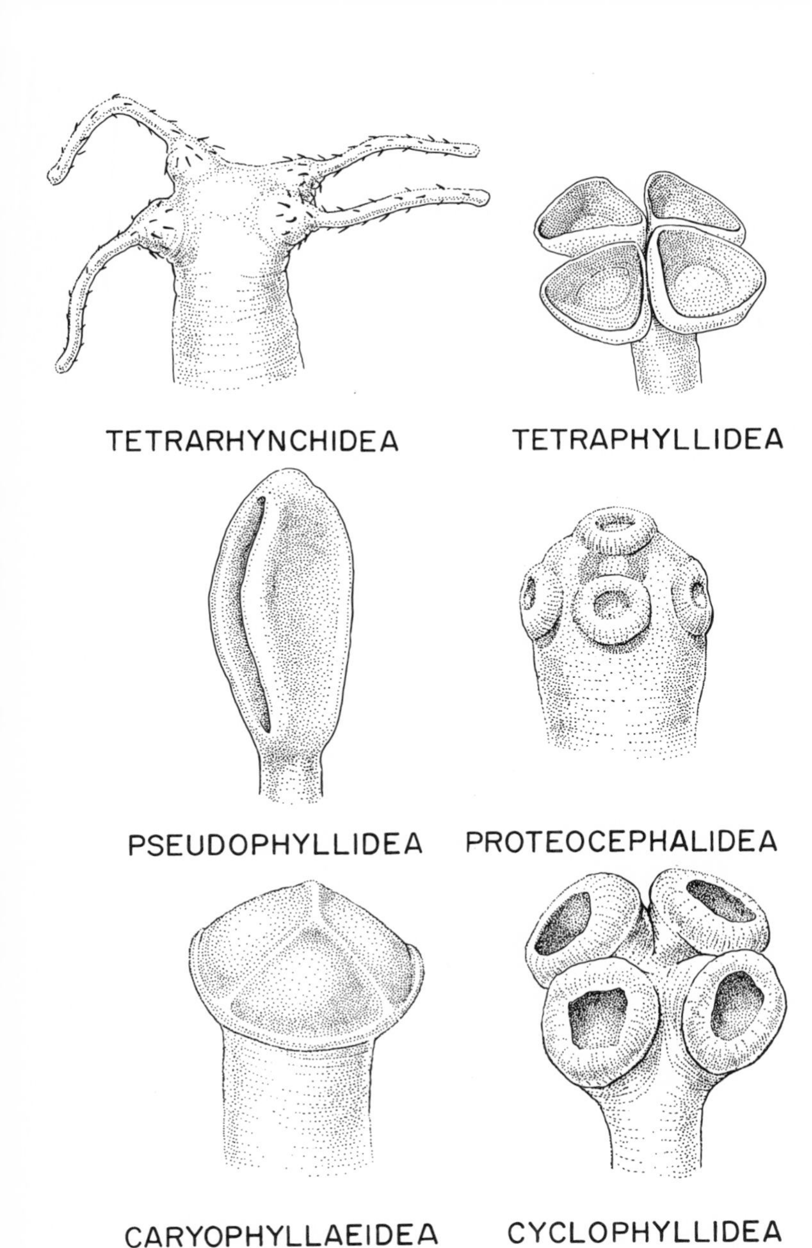

Figure 1. Holdfasts of six major orders of cestodes.

Ordinal Assignment of Families

The assignment of the family Mesocestoididae to the order Cyclophyllidea is an excellent example of undue reliance on adult structural characters. The criterion for this grouping was the presence of four suckers on the scolex; this is indeed a character of cyclophyllids, but also of adult proteocephalids. Mesocestoidid adults, however, possess a medially situated genital pore which is neither a cyclophyllid nor a proteocephalid character. On the basis of negative experimental evidence, it was long suspected that *Mesocestoides* required two intermediate hosts for the completion of the life cycle. Although this was never demonstrated, feeding of gravid proglottides containing tetrathyridial preadult stages in nature to rodents, lizards, and other animals, never resulted in infection. A three-host cycle is characteristic of several cestode orders but not of the Cyclophyllidea. Thus, in spite of contradictory evidence, the family Mesocestoididae has remained in the order Cyclophyllidea and this grouping has not been seriously questioned by contemporary taxonomists.

Recent studies on *Mesocestoides* have shown several new facts which, in my opinion, represent additional evidence for the present misplacement of this family. The structure of the developmental stages from the hatched oncosphere to the tetrathyridium obtained by culture *in vitro* (Voge, 1967) shows resemblance to proteocephalid procercoids. There is an apical organ, development and loss of a "cercomer," and the subsequent loss of the apical organ concomitant with the formation of suckers. It has been shown further that the requirements of the organism change as transformation from the procercoid-like stage to the tetrathyridium is about to occur (Voge and Seidel, 1968). Specifically, whole blood (for which hemoglobin can be substituted) is a requirement for full development to the tetrathyridial stage which normally occurs in the tissues and body cavities of vertebrates. This indicates that transition from one larval stage to the next is, in nature, dependent upon passage to a vertebrate host. Also, the procercoid-like stages are, *in vitro* at least, temperature sensitive and do not tolerate temperatures above 30°C. The tetrathyridial stages grow well at 37°C or at lower temperatures.

Figure 2 illustrates schematically the life cycle of *Mesocestoides corti* as far as it is known in nature. The first intermediate host, possibly a free-living mite (Soldatova, 1944) has not been unequivocally determined.

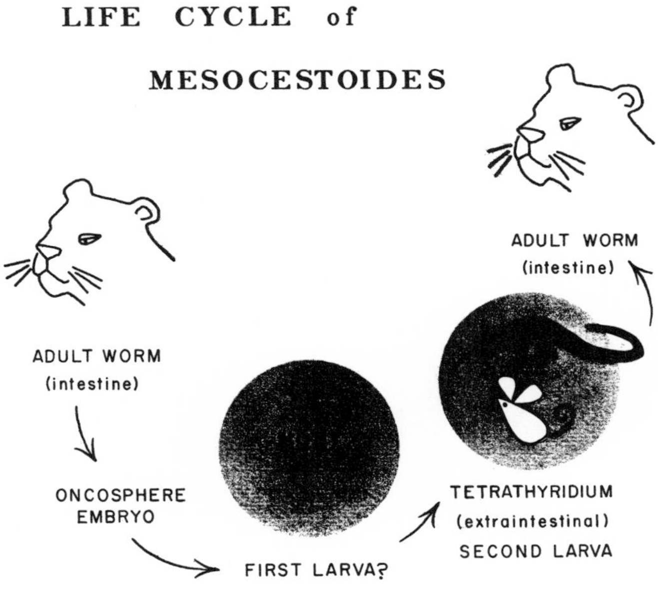

Figure 2. Diagram of life cycle of *Mesocestoides corti* as far as known to occur in nature.

Figure 3 illustrates the developmental pattern of *Mesocestoides corti* as established experimentally in the laboratory. An unusual feature, of this species at least, is the ability to reproduce asexually at the tetrathyridial stage (Specht and Voge, 1965). This multiplication is initiated by the formation of supernumerary suckers followed by longitudinal splitting into two individuals, a type of multiplication heretofore not found in cestodes. Moreover, upon entry

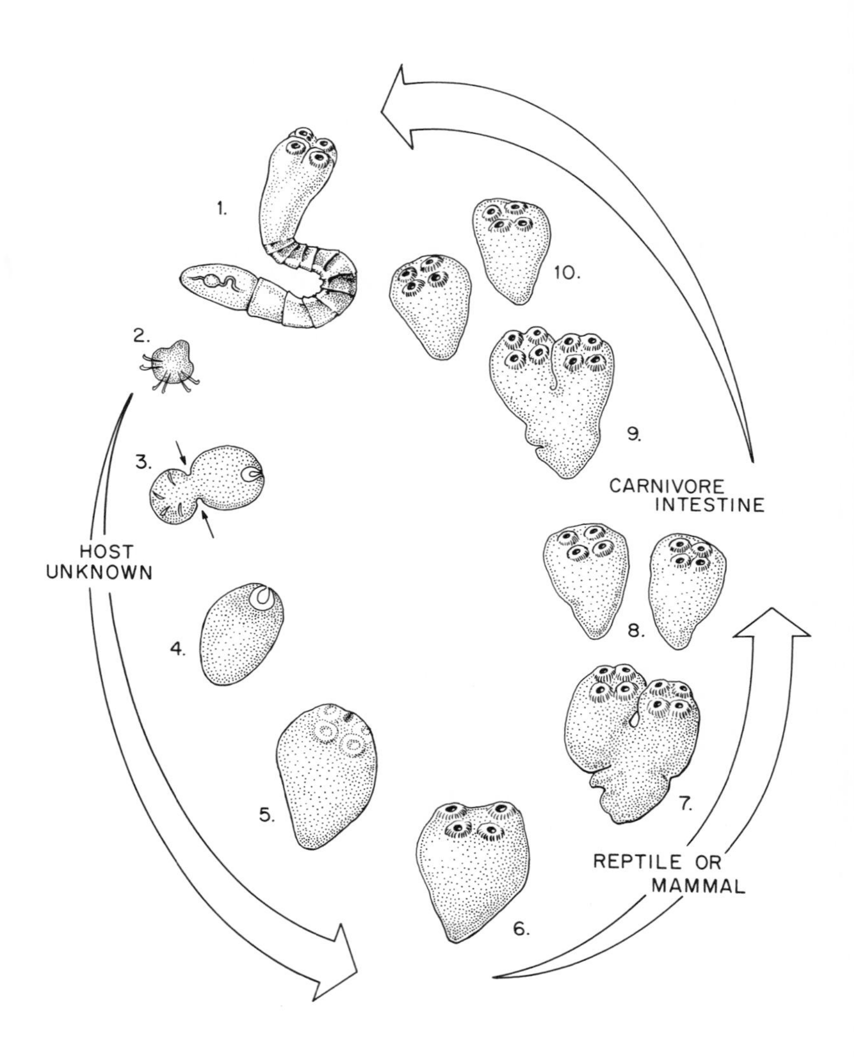

Figure 3. Developmental sequence of *Mesocestoides corti* illustrating early developmental stages *(2–5)*, tetrathyridium and asexual multiplication in second intermediate host *(6–8)*, and asexual multiplication with subsequent formation of adult worms in intestine of definitive host *(9–10)*. Not illustrated here is potential reinvasion of tissues from intestinal lumen of carnivore with continuing asexual multiplication of tetrathyridial stage.

of tetrathyridia into the definitive host (dogs, skunks) asexual multiplication continues in the small intestine for several weeks before transition to the adult takes place (Eckert *et al.*, 1969). This capacity, thus far also unique among cestodes, eventually results in very large numbers of adult worms in the definitive host. Heavy infections with adult *Mesocestoides* are commonly encountered under natural conditions. Whether or not the ability to multiply asexually is limited to certain species only, remains to be determined.

Differentiation of species of *Mesocestoides* has been based on size relationships and structure of adult worms only. Unfortunately, the adults look very much alike and the characters of many of the so-called species in North America overlap considerably (Voge, 1955). It is most likely that life history data and growth characteristics will be the only sound means of delimiting the species in this genus.

The above example of the family Mesocestoididae serves to illustrate two major points. The first is the very tenuous affinity of the group with the order Cyclophyllidea, which should have been suspected on the basis of adult structure and bionomics. Second, the structural similarity among adult worms shows the necessity of using additional criteria for species differentiation in this genus. There now remains the problem of deciding where the family Mesocestoididae really belongs, and this in turn requires consideration of what constitutes or should constitute an order. The Cyclophyllidea is a very heterogeneous assemblage and, depending on the "splitting or lumping" propensities of individual taxonomists, many of the Cestoda might profitably be reassigned or subdivided, as long as some consistency in procedure is adopted. However, a disregard of bionomics, development, and growth patterns of stages other than the adult, is biologically unsound.

Grouping of Subfamilies

The assignment of diverse groups to one and the same family is exemplified by the Anoplocephalidae which, among others, contains the subfamily Linstowiinae (see Joyeux and

Baer, 1961; Yamaguti, 1959). Spassky (1951) raised the Linstowiinae to family rank and created the superfamily Anoplocephaloidea to include the Linstowiidae. Milleman and Read (1953) also gave evidence for the desirability of elevating the Linstowiinae to family rank.

The only character which members of the Anoplocephalidae have in common is a scolex with four suckers, without a rostellum or hooks. In all other respects the different subfamilies vary among each other in several basic and important characters. Comparison of the subfamilies Anoplocephalinae and Linstowiinae shows that, on the basis of available data, there exist major differences in egg structure, intermediate hosts, and morphology of the larval stages. In the anoplocephalines investigated, the larval stages are cysticercoids similar in overall appearance to those of *Hymenolepis* (see Stunkard, 1941); in the linstowiine genus *Oochoristica* the "cysticercoid" is actually a young preadult without specialized and deciduous envelopes. Larval stages of anoplocephalines develop in free-living mites; those of *Oochoristica* and *Atriotaenia* develop in insects (Gallati, 1959; Millemann, 1955; Hickman, 1963; Widmer and Olsen, 1967). The definitive hosts of the Anoplocephalinae are primarily marsupials, rodents, rabbits, herbivores, primates, and birds; those of the Linstowiinae are reptiles, marsupials, rodents, carnivores, and bats.

Figure 4 illustrates the postembryonic developmental pattern of *Oochoristica* in insects; Figure 5 shows the developmental stages of anoplocephalines (*Cittotaenia, Bertiella*). Obvious differences between the two patterns include presence or absence of tail formation and of the deciduous tissues surrounding the scolex in fully developed preadults. The eggs of many anoplocephalines contain a "pyriform apparatus" (Fig. 5) while those of linstowiines do not. Although the number of known life histories in either group is relatively small, the known developmental patterns differ sufficiently to furnish evidence of distinctness.

One of the major problems is lack of information about the postembryonic development in many other anoplocephalid genera. Development and larval structure obviously cannot be the sole criteria for the separation of taxonomic

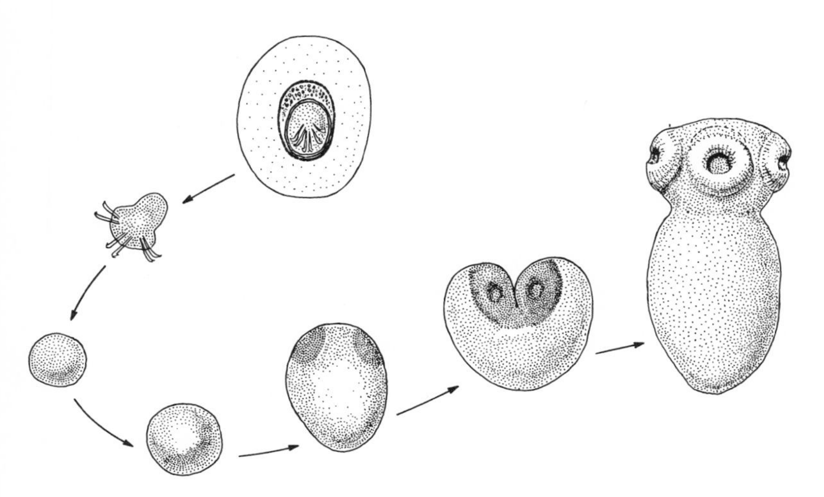

Figure 4. Diagram of developmental pattern of *Oochoristica* from oncosphere to infective preadult (*adapted from various sources*).

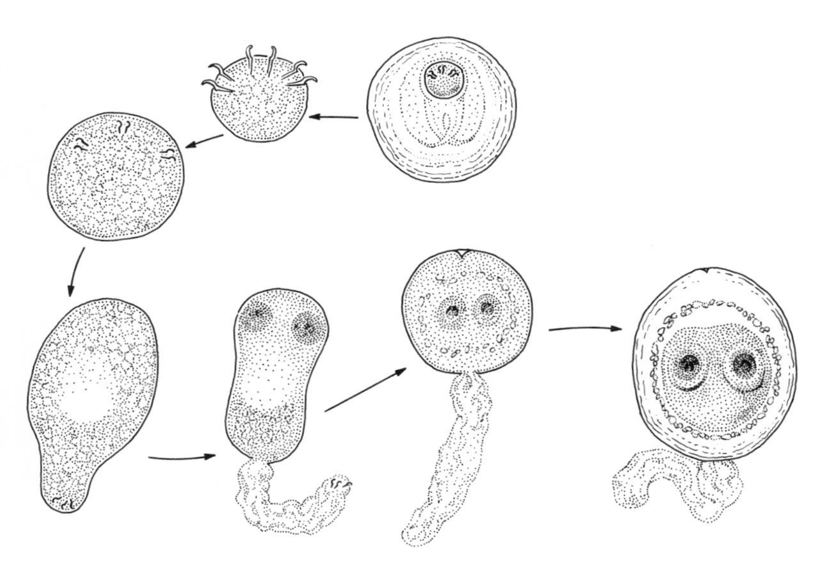

Figure 5. Diagram of developmental sequence of anoplocephalines from oncosphere to cysticercoid (*adapted from Stunkard, 1941*).

categories, as there are frequently similarities in these respects among divergent groups. The importance placed on any one characteristic is largely a matter of individual judgment. A whole complex of stable characters, however, should provide sound evidence for differentiation and distinctness between groups and should not be neglected in favor of conformity to the past.

The family Dilepididae is probably one of the most haphazardly assembled families among the cestodes. There is no single character which the many genera assigned to this family have in common (See Yamaguti, 1958, p. 227). Division into the three subfamilies (Dilepidinae, Dipylidiinae, Paruterininae) does not appreciably alleviate the difficulties encountered in keying out genera or species. The disparity in structure among the genera assigned to the subfamily Paruterininae for example, makes one wonder what possible reason made taxonomists throw these forms into one "basket."

Although all the genera of this subfamily contain a paruterine organ, this structure varies considerably in the different genera, and also occurs in nondilepidid cestodes; all other morphological features (scolex, gonads, and so on) are also dissimilar. Furthermore, the life histories and larval stages differ markedly, as species of *Paruterina* have rodent intermediate hosts with plerocercoid-like larvae (Freeman, 1957), while those of *Metroliasthes* have insects as intermediate hosts (Jones, 1936) and larvae very different from *Paruterina*. A third genus in this subfamily, *Anoncotaenia,* has embryos which do not in any way resemble the "usual" six-hooked oncosphere but rather have the shape of a slender, elongate vermicule which apparently does not bear any hooks (refer to the article by Joyeux and Baer, 1961, p. 446).

How this assemblage is supposed to relate to one and the same family, Dilepididae, which within its other subfamilies contains a comparably wide array of forms, is indeed obscure. Neither the criteria of practicality nor those of biological kinship or origin appear to have been met by this grouping.

Definition of Species

The inadequacy of using only certain adult structural characters in the definition of a species is exemplified by the unarmed hymenolepidids from rodents, particularly *Hymenolepis citelli* and *H. diminuta*. The adult stages are so similar that they were considered by some taxonomists to be but one species. Figures 6 and 7 show mature proglottides of *H. diminuta* and *H. citelli*. Size and position of internal organs do not provide a reliable means of distinguishing between them. Variability patterns in the position of the testes differ in the two forms (Voge, 1952) and were considered to be evidence of specific distinctness. In nature, *H. citelli* is typically found in squirrels and occasionally in other rodents, while *H. diminuta* occurs predominantly in Norway rats or Brown rats. Neither species is strictly host-specific. Among laboratory animals *H. citelli* grows best in hamsters; the rat is not a satisfactory host. *H. diminuta* grows well in both rats and hamsters. The cysticercoids of both species develop in tenebrionid and other types of beetles. However, study of hatching and postoncospheral development showed the following: Hatching *in vitro* of oncospheres of *H. citelli* proceeds optimally in the presence of trypsin, while trypsin combined with amylase is optimal for *H. diminuta*. Amylase does not affect any of the embryonic membranes of *H. citelli*, but does so in *H. diminuta* (Berntzen and Voge, 1965). Development of cysticercoids is completed in 11 days in *H. citelli* and in eight days in *H. diminuta* (in *Tribolium confusum* at 30°C) (Voge and Turner, 1956). Fully developed cysticercoids of both species are shown in Figures 8 and 9. The differences in structure of the tissues surrounding the scolex are readily apparent.

Surface views of the eggshell of both species photographed at the same magnification with the scanning electron microscope (Cambridge, Mark II) show distinct differences in size of the external protuberances, these being much coarser in *H. diminuta* (Fig. 10) than in *H. citelli* (Fig. 11). The thickness of the shell also differs, being

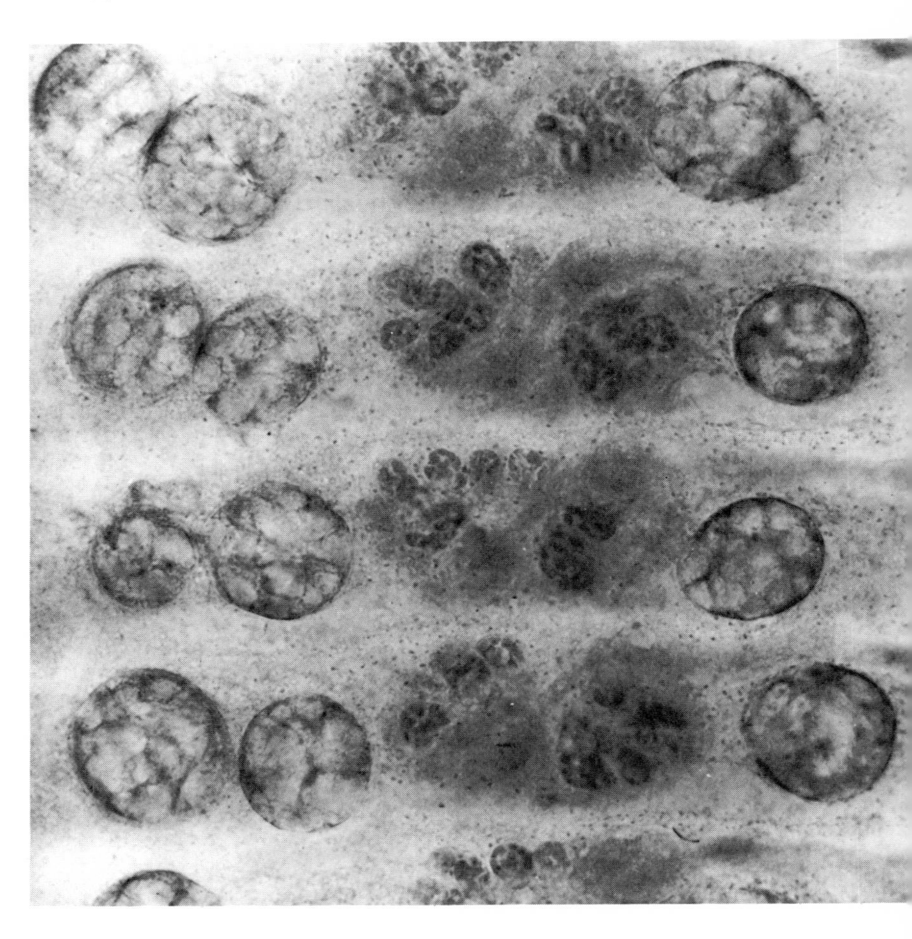

Figure 6. Mature proglottides of *Hymenolepis diminuta* (*from Voge, 1952; Courtesy University of California Press*).

about twice as thick in *H. diminuta* as in *H. citelli*. Furthermore, the eggs of the two species differ in the structure of the vitelline membrane which is much smaller in diameter in *H. citelli* than in *H. diminuta*. The oncosphere coat in *H. citelli* always lacks the polar thickenings often observed in *H. diminuta*.

While scanning electron microscopy is not a very practical means of differentiation between species, some differences in eggshell patterns and membranes are readily revealed by studies with the light microscope. It is most likely that the structure of the eggshell, and of some of the other membranes enclosing the embryo, is generally stable, and should be useful in the differentiation of some species.

Thus it is evident that the developmental and structural characters of preadult stages are much more distinct and

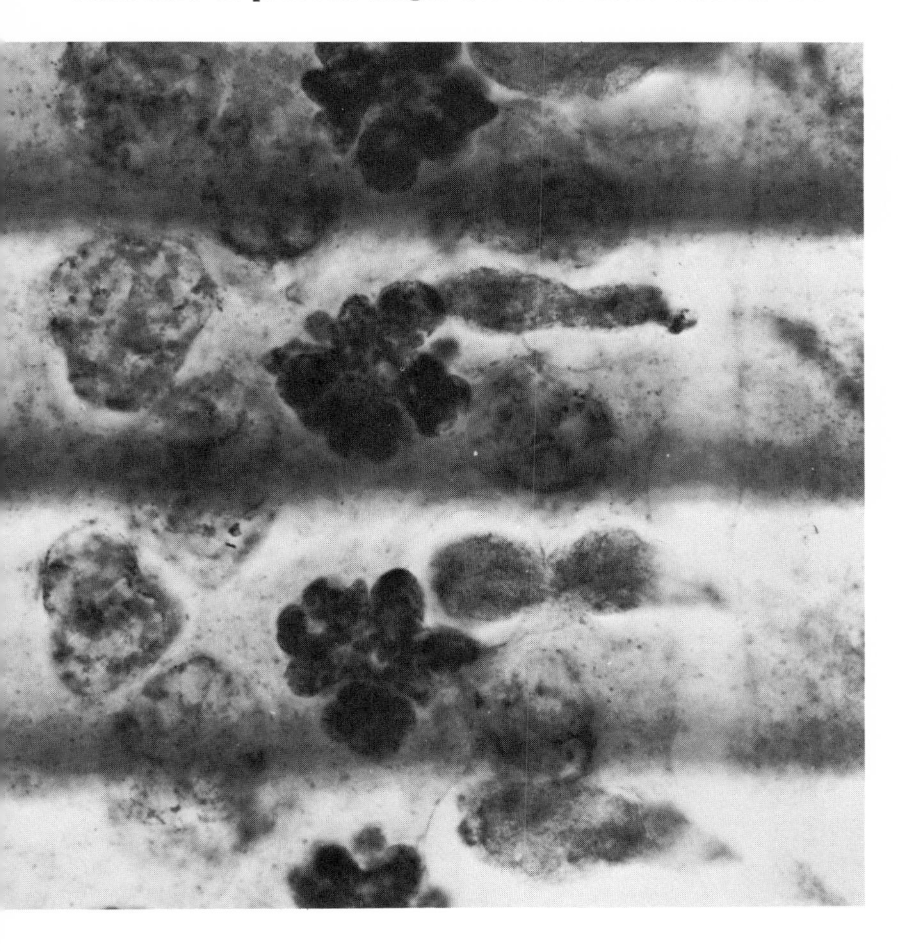

Figure 7. Mature proglottides of *Hymenolepis citelli* (*from Voge, 1952; Courtesy University of California Press*).

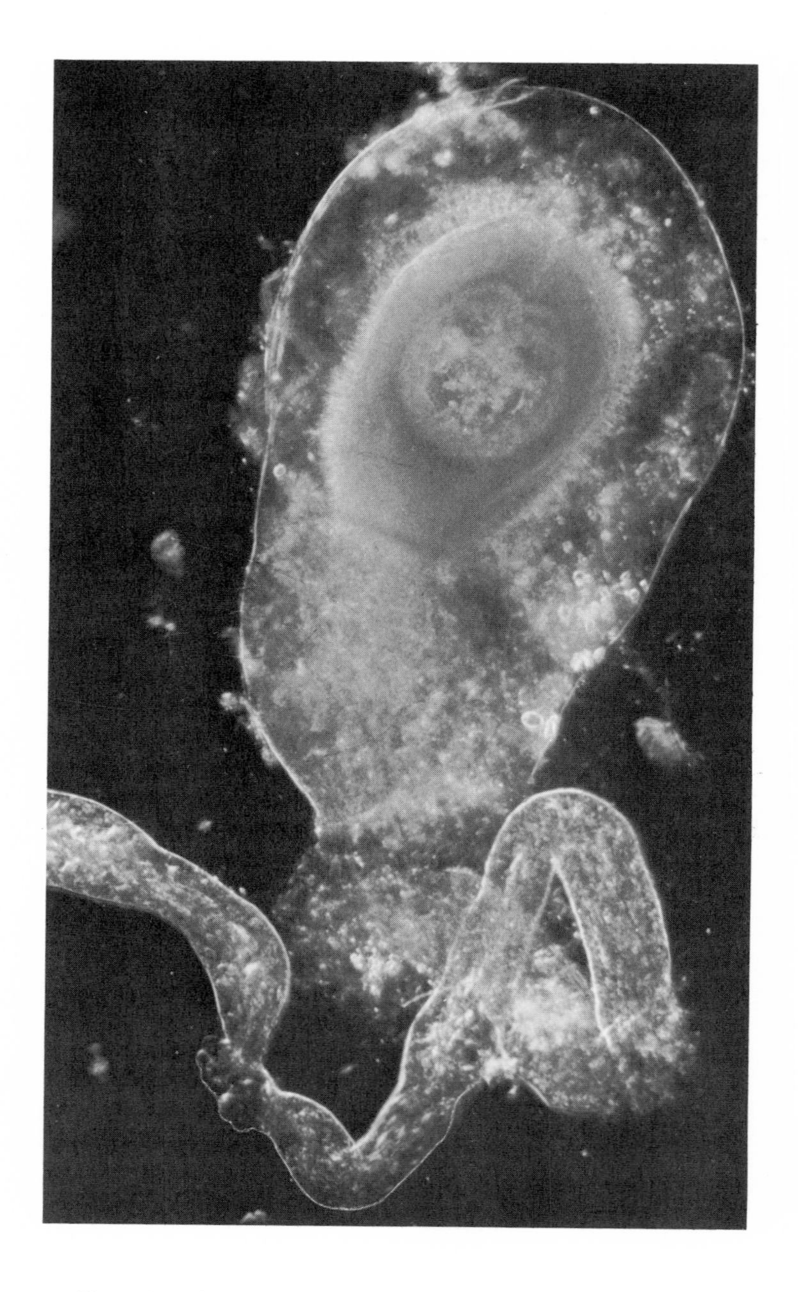

Figure 8. Fully developed cysticercoid of *Hymenolepis diminuta*. (*Photograph by Zane Price.*)

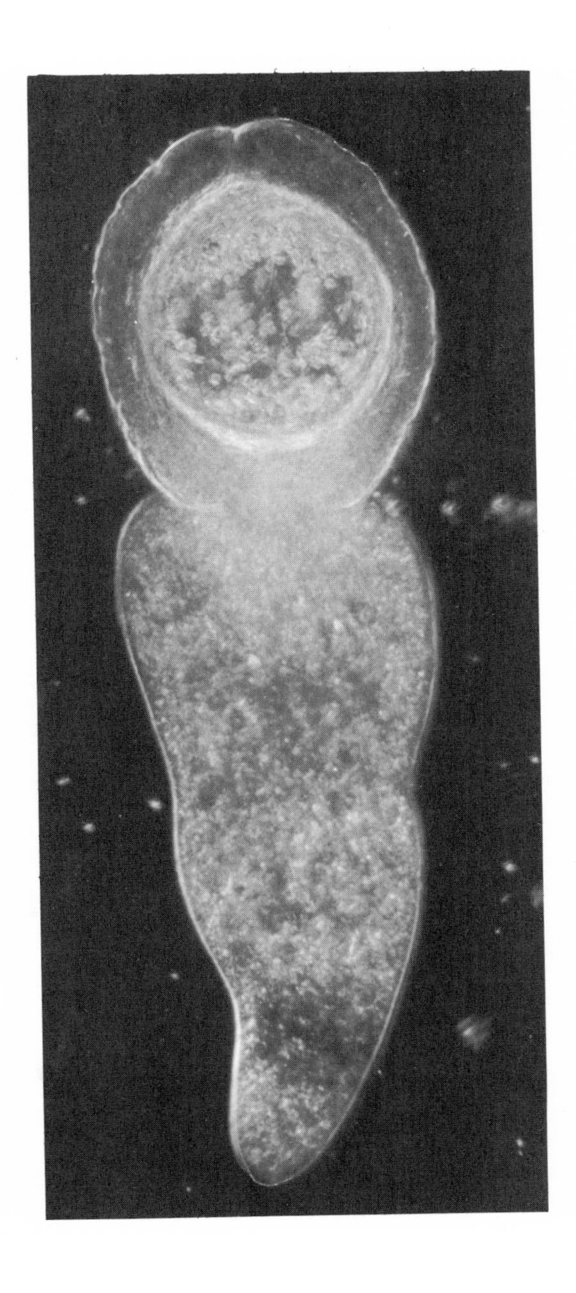

Figure 9. Fully developed cysticercoid of *Hymenolepis citelli*. (*Photograph by Zane Price.*)

useful for differentiation of *H. citelli* from *H. diminuta* than is the structure of the adults (Voge, 1960). Similar examples may be found in other cestodes.

In the taxonomy of other invertebrates, particularly the Arthropoda, larval morphology and developmental criteria are used extensively. In mosquitoes, bionomics and structure of preadult stages represent an important and even essential part of the description of a species. Whenever available, data on the structure of larval stages and the pattern of their development could be profitably included in the description of cestode species. One of the difficulties is, of course, the lack of definition of our terms. For example, the terms "cysticercoid," "procercoid," and "plerocercoid" are poorly characterized and are at times used indiscriminately, so that precise differentiation between some larval types has either not been made accurately, or has not been

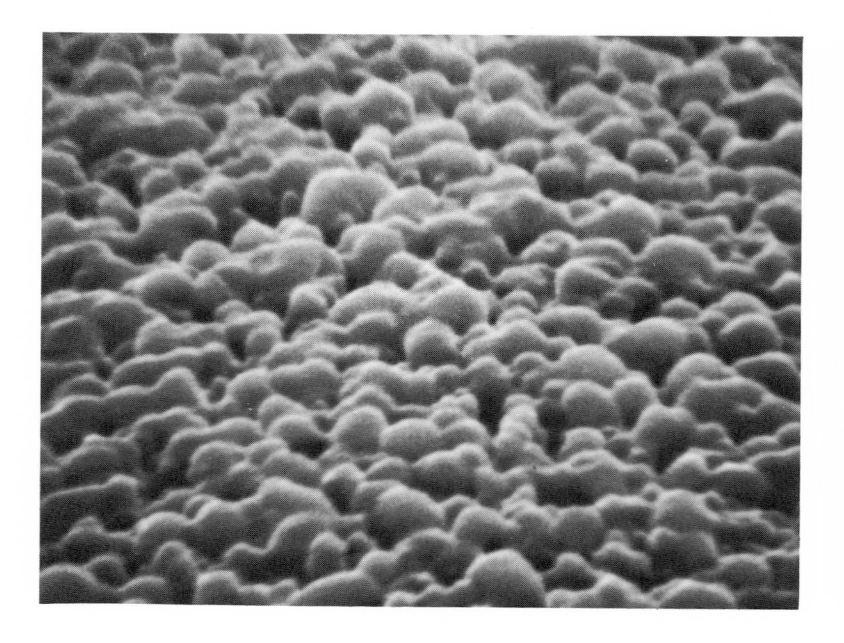

Figure 10. Photograph of surface of the eggshell of *Hymenolepis diminuta*; × 12,750. (*Courtesy California Institute of Technology.*)

made at all. As illustrated by the examples of *Oochoristica* and anoplocephalines, the small preadult of *Oochoristica* or *Atriotaenia* surely cannot be called a cysticercoid as is the larval stage in anoplocephalines or *Hymenolepis*. In the latter, the preadult is contained in specialized deciduous envelopes and, depending on the species, a tail may or may not be present. In *Oochoristica* there are no such structures and the organism withdraws into itself.

One of our most immediate needs is for a redefinition of the terms employed in the designation of preadult developmental stages. Sufficient description of structure and development must be included to permit distinction between larval types, to clarify the similarities between them, and to make possible the recognition of relationships and intergrading forms, when discovered.

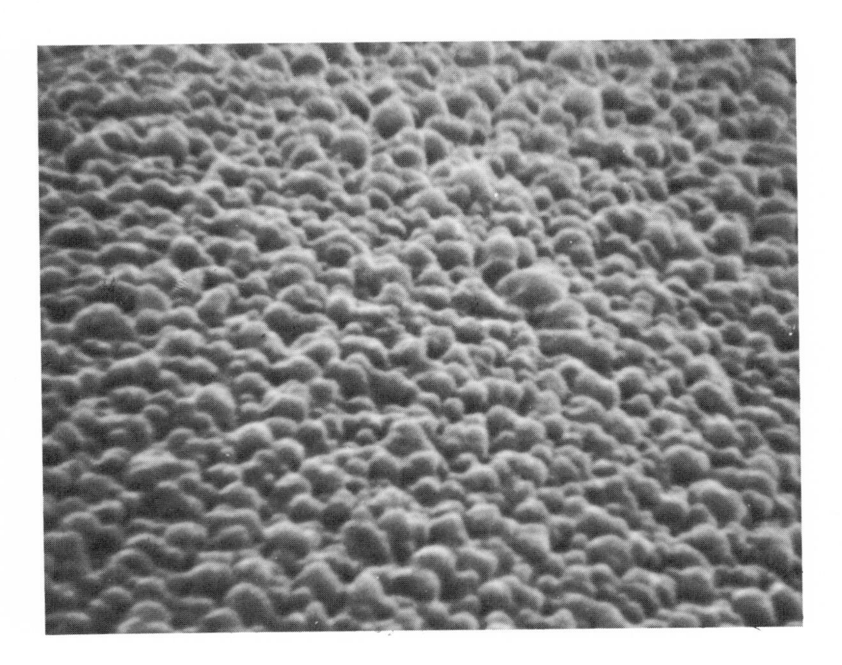

Figure 11. Photograph of surface of the eggshell of *Hymenolepis citelli*; × 12,750. (*Courtesy California Institute of Technology*.)

Present Problems and Future Systematics

One of the most pressing problems involving cestode systematics (and systematics in general) is expressed by Sokal (1964), who states that data of value in taxonomy accumulate at an ever increasing rate and that by present procedures this information cannot be digested and processed. Also, although most of taxonomy has its basis in morphology, research in biochemistry, physiology, behavior, and other disciplines has produced an impressive array of characters for evaluation of taxonomic affinities; this information is clearly pertinent to classification.

What type of information, other than morphological, is presently available for cestodes, and how reliable or useful might such data be for the systematics of these organisms? In the preceding pages I briefly discussed the importance of life histories, developmental patterns, larval structures, and some neglected aspects of adult morphology. Anyone working in the field of cestode taxonomy knows that the usefulness of a given structure or characteristic for the differentiation between species or groups of species may vary considerably among different groups of cestodes. In like manner, we cannot expect that any one biochemical attribute will serve in all instances to delimit one species from another in clear fashion, even if all the necessary data were at hand. As stated by Florkin and Mason (1960) : "the extent of biochemical diversity . . . can only be determined by study of all living organisms."

Any stable characteristic of an organism can be a useful criterion in systematics if the systematist is sufficiently well trained to apply it with discrimination and caution. This brings us to the problem of systematists and taxonomists who, at present, are a rapidly vanishing race in parasitology.

Aside from the fact that a morphologically oriented systematist will be at a distinct disadvantage when competing for jobs, the modern biology curriculum has shifted in emphasis toward the chemical and physical sciences, so much so that this frequently results in inferior or insufficient training in other, equally important, branches of biology. The essential educational compromise and balance between structure and function of organisms is yet to be established. Meanwhile, the scarcity of systematists and taxonomists presents a real difficulty for those who require identification of species of medical, veterinary, or economic importance. The functioning of many of our governmental laboratories concerned with agriculture, fish and wildlife, and the control of human, animal, and plant diseases, is dependent upon the accurate identification of pathogens. This task can be accomplished only with thorough training in morphology and the techniques pertaining to the observation, preparation, and collection of organisms. It is to be expected that the lack of suitably trained personnel will soon result in a crisis unless a more reasonable attitude or a sound compromise is achieved by those who govern our educational policies.

As stated by Sokal (1964), the new systematist will have to be trained much more broadly and his training will have to include biochemical techniques, in addition to other essential biological procedures. Also, taxonomy will have to move into the field of electronic data processing and the new museum will have to have access to a sophisticated computer which will contain morphological, biochemical, and other pertinent information.

I imagine that in the not too distant future the diagnosis and description of a species of cestode will contain the information tabulated on page 70.

The brave new taxonomists will have to cope with data such as these in years to come. It is to be hoped that they will improve on our present-day efforts to create some sort of order without unduly distorting the facts.

Species: X_y

Specific diagnosis:

1. *Morphology:* Eggs, embryos, larvae, adults
2. *Development:* Hatching, hosts, speed (larvae, adults), excystment triggers, sensitivity to environment factors
3. *Biochemistry:* Metabolic pathways (all stages). Energy sources, proteins, and amino acid sequences
4. *Genetics:* DNA hybridization data, karyotypes
5. *Variability:* Morphological and biochemical, in relation to environment; for example, total capacity
6. *Behavior:* Responsiveness to stimuli

Pertinent illustrations

Summary of essential characteristics and index of relatedness to other species in genus

REFERENCES

Berntzen, A. K., and M. Voge. 1965. *In vitro* hatching of oncospheres of four hymenolepidid cestodes. *J. Parasitol.* 51:235–242.

Brand, T. von, M. U. Nylen, N. G. Martin, and F. K. Churchwell. 1967. Composition and crystallization patterns of calcareous corpuscles of cestodes grown in different classes of hosts. *J. Parasitol.* 53:683–687.

Eckert, J., T. von Brand, and M. Voge. 1969. Asexual multiplication of *Mesocestoides corti* (Cestoda) in the intestine of dogs and skunks. *J. Parasitol.* 55:241–249.

Florkin, M., and H. S. Mason. 1960. *Comparative Biochemistry.* Vol. I. Academic Press, New York.

Freeman, R. S. 1957. Life cycle and morphology of *Paruterina rauschi* n. sp. and *P. candelabraria* (Goeze, 1782) (Cestoda) from owls, and significance of plerocercoids in the order Cyclophyllidea. *Canad. J. Zool.* 35:349–370.

Gallati, W. W. 1959. Life history, morphology and taxonomy of *Atriotaenia* (*Ershovia*) *procyonis* (Cestoda: Linstowiidae), a parasite of the raccoon. *J. Parasitol.* 45:363–377.

Hickman, J. L. 1963. The biology of *Oochoristica vacuolata* Hickman (Cestoda). *Proc. R. Soc. Tasmania* 97:81–104.

Jones, M. F. 1936. *Metroliasthes lucida,* a cestode of galliform birds in arthropod and avian hosts. *Proc. Helm. Soc. Wash.* 3:26–30.

Joyeux, C., and J. G. Baer. 1961. Classe des cestodes. In: P. P. Grassé, *Traité de Zoologie,* Vol. IV. Masson et Cie, Paris.

Millemann, R. E. 1955. Studies on the life history and biology of *Oochoristica deserti* n. sp. (Cestoda: Linstowiidae) from desert rodents. *J. Parasitol.* 41:424–440.

——— and C. P. Read. 1953. The biology of *Oochoristica* and the status of Linstowiine cestodes. *J. Parasitol.* 39:29 (Suppl.).

Mueller, J. F. 1965. Further studies on parasitic obesity in mice, deer mice and hamsters. *J. Parasitol.* 51:523–531.

——— 1968. Growth stimulating effect of experimental sparganosis in thyroidectomized and hypophysectomized rats, and comparative activity of different species of *Spirometra. J. Parasitol.* 54: 795–801.

Read, C. P. 1968. Intermediary metabolism of flatworms. In: Florkin, M. and B. T. Sheer, Eds., *Chemical Zoology,* Vol. II. Academic Press, New York.

——— L. T. Douglas, and J. E. Simmons. 1959. Urea and osmotic properties of tapeworms from elasmobranchs. *Expl Parasitol.* 8:58–75.

——— and J. E. Simmons. 1963. Biochemistry and physiology of tapeworms. *Physiol. Rev.* 43:263–305.

——— and M. Voge. 1954. The size attained by *Hymenolepis diminuta* in different host species. *J. Parasitol.* 40:88–89.

Rothman, A. H. 1959. Studies on the excystment of tapeworms. *Expl Parasitol.* 8:336–364.

Simmons, J. E. 1961. Urease activity in trypanorhynch cestodes. *Biol. Bull.* 121:535–546.

Sokal, R. R. 1964. The future systematics. In: C. A. Leone, Ed. *Taxonomic Biochemistry and Serology,* pp. 34–48. Ronald Press, New York.

Soldatova, A. P. 1944. A contribution to the study of the development cycle in the cestode *Mesocestoides lineatus* (Goeze, 1782), parasitic of carnivorous mammals. *C. R. (Doklady) Acad. Sci. U.R.S.S.* 45:310–312.

Spassky, A. A. 1951. *Osnovy cestodologii.* Vol. I. *Anoplocephaliati, Lentochnie gelminti domaschnih i dikih zhivotnih.* Akademia Nauk, Moscow.

Specht, D., and M. Voge. 1965. Asexual multiplication of *Mesocestoides* tetrathyridia in laboratory animals. *J. Parasitol.* 51: 268–272.

Stunkard, H. W. 1941. Studies on the life histories of the anoplocephaline cestodes of hares and rabbits. *J. Parasitol.* 27:299–325.

——— 1953. Life histories and systematics of parasitic worms. *Syst. Zool.* 2:7–18.

Voge, M. 1952. Variation in some unarmed hymenolepididae (Cestoda) from rodents. *U. Calif. Publ. Zool.* 57:1–52.

——— 1955. North American cestodes of the genus *Mesocestoides*. *U. Calif. Publ. Zool.* 59:125–156.

——— 1960. Studies in cysticercoid histology. V. Observations on the fully developed cysticercoid of *Hymenolepis citelli*. *Proc. Helminth. Soc. Wash.* 28:1–3.

——— 1967. Development *in vitro* of *Mesocestoides* (Cestoda) from oncosphere to young tetrathyridium. *J. Parasitol.* 53:78–82.

——— and Graiwer, M. 1964. Development of oncospheres of *Hymenolepis diminuta,* hatched in vivo and in vitro, in the larvae of *Tenebrio molitor*. *J. Parasitol.* 50:267–270.

——— and Seidel, J. S. 1968. Continuous growth *in vitro* of *Mesocestoides* (Cestoda) from oncosphere to fully developed tetrathyridium. *J. Parasitol.* 54:269–271.

——— and Turner, J. A. 1956. Effect of temperature on larval development of the cestode, *Hymenolepis diminuta*. *Expl Parasitol.* 5:580–586.

Widmer, E. A., and Olsen, O. W. 1967. The life history of *Oochoristica osheroffi* Meggitt, 1934 (Cyclophyllidea: Anoplocephalidae). *J. Parasitol.* 53:343–349.

Yamaguti, S. 1959. *Systema Helminthum*. Vol. II. *The Cestodes of Vertebrates*. Interscience, New York.

Discussion

Question: When is this new era of taxonomy and systematics supposed to come about?

Dr. Voge: Maybe in two hundred years we will be close.

Question: I think it is understood by everyone that with the advanced techniques coming into use today we will find completely new animals as we begin to explore the ocean depths. And these animals will have completely new parasites in them. I think Dr. Manter would agree that this is particularly true for parasites of elasmobranchs and deep sea teleosts which have not been looked at to any degree. I think one of the problems taxonomists will have to cope with is the conservative attitude of the present-day workers who have established these sound orders that everyone has gone by.

Dr. Voge: I think this is very true. There is a great deal yet to be discovered.

Question: Dr. Voge, regarding the Taeniidae, do you think it is valid to consider various genera as separate on the basis of the form of the cysticercus, such as *Multiceps, Hydatigera,* and so on?

Dr. Voge: I am not sure about the Taeniidae. If we should separate the Cyclophyllidea into several orders—and as I said there might be good reasons to do this—I would put the taeniids into a separate one; this has been suggested before. I am not sure of the weight that should be placed on differences in larval stages in each case. This is something that would have to be looked at very carefully, of course. I do not know if these variations in larval stages should constitute generic differences or not; I think they might. I would have to think about it more.

John L. Crites, Professor of Zoology at Ohio State University, has published for the past several years on the biology and systematics of parasitic nematodes, both from vertebrates and invertebrates. He has served as Agency for International Development–National Science Foundation consultant at Aligarh Muslim University and Banaras Hindu University in India, and is Professor of Parasitology at Stone Laboratory of Hydrobiology.

John L. Crites/Problems in Systematics of Parasitic Nematodes

The problems in systematics of nematodes parallel those of the entire field of systematics: description, identification, classification, integration, synthesis, and communication. It is estimated that at the present time about 10,000 species of nematodes have been named; this is surely only a small fraction of the whole. Hyman (1951) estimated that there may be as many as 500,000 species of nematodes. Obviously nematologists still have a great task to equate groups of nematodes to the species level and to describe them in a usable manner. Some biologists might question whether this is a problem or a challenge. I prefer to think of it as the latter.

Description

If one studies descriptions of new species of nematodes named during the last ten years what does he find? The nematologist has recognized that he is working with populations. He studies the variations represented by individuals in a group and records these in his written descriptions and in his illustrations. He designates one of the specimens as the *type* in order to avoid confusion about the name. These descriptions are primarily morphological although they usually contain data about habitat of the nematodes or the host, sex ratio, numbers found, juvenile stages, and locality. Occasionally there is a description in which mention is made of the fact that some males and females were found in copulation; usually no other information is given or known about interbreeding, although it is usually assumed until proved otherwise. The nematological taxonomist describing a new species is not a speciationist. He is actually a biological reporter who reports only what he finds.

Some biologists have claimed that this kind of description creates problems. The claim is that these descriptions are not biological, that interbreeding and reproductive isolation have not been demonstrated for these groups. It is true that these facts have not been used, to my knowledge, in the description of any parasitic nematodes. The parasitologist rarely finds himself in direct possession of data on breeding populations, barriers, or genetic mechanisms in sufficient quantity to be useful in original descriptions of new species of nematodes. The parasitologist describing new nematode groups may fail to recognize two "biological" species that are very similar on the basis of comparative data. He may also recognize, as two species, groups which vary morphologically but which may later be shown to interbreed. He does this not because he does not believe that these species evolved. He realizes that they are a living product of nature; indeed, his work is an attempt to reflect in the easiest and most practical way the natural situation. The fact is that interbreeding and outbreeding studies are almost always carried out with previously described and identified groups of animals. Some nematodes are monoecious and others have asexual generations and will not fit the "biological species concept" in any case. Pioneering original descriptions based on comparative demonstrable differences are necessary to systematics in nematology.

If there is a problem here it is financial support for this kind of descriptive work with recognition of the fact that it needs to be done. Descriptions and monographs of high quality constitute a contribution both to the systematist and the nonsystematist. Descriptions are necessary before there can be identification, and taxonomic identification precedes most other biological studies.

Identification

Identification is the proper placement of a specimen or a group of specimens into the established order of the taxonomic hierarchy.

Are there problems of identification unique to parasitic nematodes? Parasitology has been described as a study of the host-parasite relationship. The host provides added behavioral, ecological, distributional, and evolutionary factors. Wallace (1962), discussing nematode ecology, points out the problems of species identity related to host specificity as they occur in plant parasitic nematodes. The role of the host relationships in intraspecific variations of nematode parasites of animals is discussed in two recent publications. M. B. Chitwood (1957) speaks of the pitfalls of variations caused by preparation techniques and then takes up the problem of taxonomic character assessment, particularly those problems pertaining to morphological variations of the same nematode species from different hosts. Haley (1962) gives an excellent review of this subject and presents an experimental viewpoint for approaching the study of this problem. I would like to quote from Haley's summary as it explains why there should be future study. He says, "Studies in systematic helminthology on the kinds, frequencies and nature of intraspecific variations, structural and otherwise, and of the influence of host relationships on them are needed for two major reasons: (1) to assess the stability of taxonomic characters used in classification, and (2) to determine the actual or potential biological significance of different variants in the colonization of new hosts and in speciation itself."

Nematologists and parasitologists have always found it a prerequisite for any other kind of biological study to have accurately identified groups of nematodes with which to work. This is fundamental to studies from the molecular to the population levels. It should be pointed out that this is a reciprocal arrangement. The systematists of nematology have a responsibility to make it possible for others to work with accurately identified species groups, but these investigations are now in turn furnishing data which can and should be used by systematists. Fortunately, in nematode systematics we can now attack the problems of identification in a more exact manner than has ever before been possible. We now have better tools with which to work and an ever increasing mass of data which must be correlated with each species group. We can now extract data from studies being conducted with nematodes in other fields such as developmental biology, biochemistry, genetics, cytology, electron microscopy, anatomy, physiology, behavior, and ecology. Blackwelder (1962) has said, "classifying the kinds of animals is generally recognized as a part of the job of systematics, but it is often overlooked that the more important and vastly more difficult part of systematics is the job of keeping track of all information discovered about each species." It is beside the point that these data are not now on a comparative basis. When we have sufficient data, which are comparative, they will be of significance to taxonomy. It may bother some of my colleagues that I bring up this matter under the heading Identification. It is my opinion that any data gathered about a species group becomes a part of the identity of that group. I suspect we will find variations in other attributes of nematode species just as we find variations in morphology. Original descriptions of nematodes which are primarily morphological will still be needed; they are, however, only a beginning and not an end in themselves. There is nothing static about species group descriptions: we can add to them or redescribe them on the basis of comparative data gained from any other field of biological study. Fortunately, it has been demonstrated many times that data from these fields usually correlate with structure at some level. The effect of these experimental or nonmorphological attributes is thus to adjust and strengthen the morphological description already in existence.

The problem, which now presents itself, is how can we handle such extensive amounts of data. If nematological systematics is to integrate all nematode biology these data must be effectively absorbed. The only answer which I can foresee is the use of data processing by means of computers. By using computers it is possible to sort, file, and retrieve all the known information about a given species of nematode. This will be no simple task and it will be costly. The fact that costs are high, or that nematologists and parasitologists do not have the use of computers in their day-to-day tasks, must not deter us from seeking the use of tools which greatly implement our studies and knowledge. After all, such obstacles have not deterred parasitologists from using electron microscopes or any other modern tools which are necessary to their work. Surely a society which maintains a computerized credit card economy will not fail to provide such services for its scientists.

A system based on all available comparative data for each identified species will lead to the discovery of useful biological correlations and differences not now evident, and it must lead to a more integrated classification.

Classification

According to Simpson (1962), the goal of classification is to produce a stable, biologically meaningful arrangement of all knowable organisms. Wharton (1959) says this about classification: "The only integrating system yet devised for handling the myriads of facts known about animals is our system of classification known as the systematic hierarchy."

The classification of nematodes has had a history more varied than most groups of animals. Nematodes have usually been recognized as a class in the hierarchy and at different times they have been placed in different phyla. Hyman (1951) gives an interesting account of the taxonomic history of these organisms. Today most nematologists recognize the nematodes as a phylum, but not all zoologists do. Hyman suggested that the nematodes be designated a class, the Class Nematoda, which she placed in the Phylum Aschelminthes. Many zoologists followed her suggestion and general zoology textbooks and textbooks of parasitology have usually used this designation. Hyman makes this statement in her discussion of the classes of the Phylum Aschelminthes: "The alternative in regard to these groups would seem to be to elevate each group to the rank of a separate phylum or to associate them loosely under one phylum.

Those who prefer the first alternative should not hesitate to adopt it." I prefer to adopt the first alternative and use the designation Phylum Nematoda. Studies on the development, life histories, biochemistry, and ultrastructure of the animals in these groups indicate that they do not have the biological affinities that their adult anatomical structures indicated. Since 1959 most textbooks concerned with plant nematology have used the category Phylum Nematoda, and some recent textbooks in invertebrate zoology have also adopted it (Meglitsch, 1967; Barnes, 1968).

It is often said that the classification of nematodes is unstable; to some extent this is true. One of the problems in their classification is the separate study of free-living nematodes, those parasitic in plants and those parasitic in animals. The reasons for this have been both academic and economic. J. R. Christie (1960) expressed his view of the situation in this way, "The line between parasitology and what we are calling nematology is not a border separating specialties of a science, but a barrier having the effect of creating two separate sciences." Later in the same paragraph he says, "I invite you to join me not in eliminating the barrier, but in reducing it to its proper status." Certainly in the classification and systematics of nematodes we must help in reducing the barrier, because very often we need to cross back and forth across the border.

The major contribution to the classification of *all* the nematodes was made by the Chitwoods in their book, *An Introduction to Nematology* (1937, 1950). This work did furnish some stability and gave us the basic framework of the classification of nematodes. In their classification the Chitwoods recognized 22 major groupings of nematodes and categorized these as superfamilies. At least 14 of these groupings contained known parasitic nematodes. Changes and disagreements have centered largely about the higher categories in this classification, specifically the raising of some superfamilies to orders. Some of these changes are now accepted and others will occur as our knowledge of both new and established taxa increases. There are those who will see these developments as a problem, but after all, while we seek stability, we do not want a rigid, static classification for a group where so much is yet to be discovered.

The Chitwoods' classification was based primarily on morphological characteristics. Most revisions of this classification, such as Yamaguti's (1961) for the classification of the nematode parasites of vertebrates, are also based for the most part on morphology. If one looks past some of the morphological diagnoses given by Yamaguti there are much better reasons for the establishment of orders, such as the Order Filariidea, than is evident from his accounts. These reasons, of course, have to do with the other attributes of these parasites, their ecology, life cycles, larval development, metabolism, and their behavior within both the intermediate and definitive hosts.

I realize that the facts gained from morphological studies are the easiest to obtain, require the least equipment, and offer a ready source of comparative data. I do not negate the importance of morphology, nor do I consider such studies second rate or classical. I recognize morphology as the foundation and beginning of many other kinds of studies. It is the basis of most available and usable taxonomic keys. Morphology will probably continue for some time to be the primary basis of classification of nematodes. However, in 1962 I said, "Every specialist has an obligation to past work in his field of endeavor, but he also has an obligation to the future. He has an obligation to make use of the data derived from modern methods." I still believe that statement. We can no longer depend almost entirely on morphology for data and evidence of relationships in the classification of nematodes. This *might* be good taxonomy, but is it good systematics? One has only to examine the current journals of parasitology and nematology to realize that the data for a more biologically integrated classification is building with each new issue. Whether or not we use computers, this new information cannot be ignored.

I am not proposing a classification which utilizes *all* attributes—parasitologists will still decide what attributes to correlate and how to interpret the significance of these attributes.

Synthesis

Synthesis, both theoretical and practical, is one of the goals of all the biological sciences. Systematics is perhaps unique in this respect for it is not just an information gathering system, it is homeostatic to the other biological sciences. It encodes, interprets, and classifies much of the information known and being discovered about organisms. This classification, in turn, gives direct and inferential information about animal attributes. This information is the basis for both theory and practice in biology from the molecular to the population levels.

Are there problems in the synthesis of data concerning parasitic nematodes? The answer, of course, is yes. They are the same problems which challenge all systematists.

Studies on interbreeding of nematode species are almost nonexistent and have had little effect on the systematics of nematodes. This kind of study—correlated with information about the zoogeography, ecology, behavior, genetics, and host-parasite relationships of nematodes—could furnish an explanation of the mechanisms of the evolution of nematode parasites. This might or might not change our classification in any noticeable way, but it is needed in our systematics.

Phylogenies are postulated, hypothetical, evolutionary sequences. Whether or not they can be utilized in classification is open to question. Phylogenies are usually constructed on existing classifications or on data furnished by these classifications. In most phylogenies the data are arranged in a certain sequence and a hypothesis is made about this sequence which, in most cases, is neither provable or unprovable. The phylogeny is theoretical.

We have had some interesting phylogenetic studies of parasitic nematodes. Notable among these are works by the following authors: Chabaud (1955*a*, 1955*b*), Dougherty (1951), Osche (1958), and Sprent (1962). Blackwelder (1962) presents the viewpoint of the conventional taxonomist concerning the misuse of phylogenies in taxonomy, and Colless (1967) presents the antiphylogeny viewpoint of a numerical taxonomist. Biochemical and immunological phylogenies are open to the same criticisms when they are postulated as a basis of classification. While we may not use these theories induced from our knowledge of animal attributes in our taxonomic studies, the question remains, do we have a place for them in systematics?

Parasitologists and nematologists have not, as yet, made much use of high speed data processing devices for taxonomic purposes. As I suggested earlier, computers could at least be utilized to sort, store, and retrieve information about nematodes. Trained parasitologists and nematologists will still be needed to interpret these data. Computers are tools, and as with all tools procedures must be developed to make the most efficient use of them. Numerical taxonomists have developed and are developing statistical procedures which they believe make the best use of these tools. The major problem for the conventional taxonomist is that the units which result from the statistical procedures of the numerical system cannot be readily integrated into the taxonomic hierarchy. OTU's are seldom communicable to other biologists. We could, I suppose, attempt to reeducate the biologists who work with nematodes to make use of this new system, but what would we do with existing literature? It seems more reasonable for us to develop statistical procedures which allow for the use of data processing equipment but which still permit the use of the tested conventional taxonomic system. Let me make it clear that I do not mean this as a condemnation of numerical taxonomy. If systematics is something more than just taxonomy then there must be room in it for experimentation. We must allow the numerical taxonomists the privilege of developing their methods and testing their results, just as they must allow the conventional taxonomists to choose what methods they will use.

The results of any method or combination of methods must be an integrated, usable, biologically meaningful systematics. The systematics of nematodes must be communicable and understandable by all biologists who need to use it.

The question is then, and perhaps this is *the* problem, has systematics now reached the stage where it can synthesize, from its own being and parts, new theories and practices which will be of value to the biological sciences and to the society which supports it? If it has, can it do this and remain completely unchanged? Dr. Schmidt said in his introduction that there had been no revolution in systematics, at least in the systematics of parasites. I believe he is correct, there has been no revolution, but we may be witnessing a small part of an evolution of both systematists and systematics. Who can say where this evolution may lead us, who can say what selection pressures there will be and what adaptations will be produced?

Systematics is not merely technical know-how; may it always be blessed with controversial interpretations.

REFERENCES

Barnes, R. D. 1968. *Invertebrate Zoology.* 2nd ed. Saunders, Philadelphia.

Blackwelder, R. E. 1962. Animal Taxonomy and the New Systematics. In: *Survey of Biological Progress,* Vol. 4, pp. 1–55. Academic Press, New York.

Chabaud, A. G. 1955*a*. Éssai d'interpretation phylétique des cycles évolutif chez les nematodes parasites de vertebrés. Conclusions taxonomiques. *Ann. Parasitol.* 30:83–126.

——— 1955*b*. Remarques sur le symétrie cephalique des nématodes et hypotheses concernant l'évolution de cette symétrie chez les phasmidiens parasites. *Bull. Soc. Zool. France.* 80:314–323.

Chitwood, M. B. 1957. Intraspecific variation in parasite nematodes. *Syst. Zool.* 6:19–23.

Chitwood, B. G. and M. B. Chitwood. 1937. *An Introduction to Nematology,* Section 1, Part 1. Monumental Printing Co., Baltimore.

——— 1950. *An Introduction to Nematology,* Section 1, Part 1. 2nd Ed. Monumental Printing Co., Baltimore.

Christie, J. R. 1960. The Role of the Nematologist. In: *Nematology: Fundamentals and Recent Advances with Emphasis on Plant Parasitic and Soil Forms,* pp. 8–11. University of North Carolina Press, Chapel Hill.

Colless, D. H. 1967. The phylogenetic fallacy. *Syst. Zool.* 16:289–295.

Crites, J. L. 1962. Morphology as a basis of identification and classification of parasites. *J. Parasitol.* 48:652–655.

Dougherty, E. C. 1951. Evolution of zooparasitic groups in the phylum nematoda, with special reference to host distribution. *J. Parasitol.* 37:353.

Haley, A. J. 1962. Role of host relationships in the systematics of helminth parasites. *J. Parasitol.* 48:671–678.

Hyman, L. H. 1951. *The Invertebrates.* Vol. 3. *Acanthocephala, Aschelminthes and Entoprocta.* McGraw-Hill, New York.

Meglitsch, P. A. 1967. *Invertebrate Zoology.* Oxford University Press, New York.

Osche, G. 1958. Beiträge zur Morphologie, Oekolegie u. Phylogenie der Ascaridoidea (Nematoda). Parallelen in der Evolution von Parasit u. Wirt. *Ztschr. Parasitenk.* 18:479–572.

Simpson, G. G. 1962. The status of the study of organisms. *Amer. Sci.* 50:36–45.

Sprent, J. F. A. 1962. The evolution of the Ascaridoidea. *J. Parasitol.* 48:818–823.

Wallace, H. R. 1962. The future of nematode ecology. *J. Parasitol.* 48:846–849.

Wharton, G. W. 1959. The future of systematic zoology. *Syst. Zool.* 8:82–87.

Yamaguti, S. 1961. *Systema Helminthum.* Vol. III. *The Nematodes of Vertebrates.* Interscience, New York.

Discussion

Question: I appreciate very much all the things that you have said and I am in sympathy with your viewpoint. I would like to ask what would happen if we today redescribe certain species and redefine them according to total present-day knowledge, as outlined by Dr. Voge? Would we be able to publish it and would it make any impact on biologists in general? If so, it might be worthwhile. The question is, would we be able to accomplish such a thing?

Dr. Crites: To answer your question, I think we still need good morphological descriptions. But I think we would have a very difficult time publishing such descriptions as you mention. I suspect we will have to develop methods to do this. At the present time the journals are likely to accept only brief morphological descriptions. I think this will gradually change, that a revolution is already taking place. Dr. Voge suggested two hundred years; I do not think it will take that long but we will slowly get to that level.

Question: In connection with this, I do think that one of our problems is that we are getting more and more species of parasites and fewer and fewer systematists, so we are forced to go to the computer. However, when one uses the computer he should pay attention to the so-called "Gigo Factor," which means "garbage in, garbage out." If you read the descriptions of a lot of species you will realize that it is "garbage in."

Question: Many people think computers mean less work. However, I recently had a student complete a thesis using numerical taxonomy, and it was very much more work than it would have been using standard techniques.

Dr. Crites: I think you are absolutely correct. That is why I said we will have to develop procedures as we go along. And I really believe we have to allow the numerical taxonomists the right to experiment with different methods to make it more simple. Right now I do not find numerical taxonomy easy to use, nor do I find OTU's to be very communicable. I think it will evolve to the point where it will be more easily used and communicable, but I cannot put any time factor on it. If we do not like what the numerical taxonomists are doing, then we should develop our own methods. It will not be easy to do, but I think it can be done.

Harold W. Manter, Emeritus Professor of Zoology at the University of Nebraska, has for many years based his research operations on the systematics and zoogeography of digenetic trematodes of fishes. He is a member of many professional and honorary societies and is past president of the American Microscopical Society, the Nebraska Academy of Sciences, and the American Society of Parasitologists.

Harold W. Manter / Problems in Systematics of Trematode Parasites

Problems in the systematics of trematodes involve difficulties in collecting, in literature, and in taxonomy proper.

Collecting and Recording

The taxonomy of trematodes involves problems of collecting because of the large number of species still unknown and because special methods of collection must be followed if specimens are to be adequate for study.

New kinds of helminths are constantly being discovered. Trematodes of marine fishes, in particular, constitute one of the least known groups of animals. Wherever extensive surveys have been made, the number of species of digenetic trematodes approach or exceed the number of species of fishes examined. Data (partly unpublished) assembled from eleven surveys show that the number of Digenea average 93 per cent of the number of fishes examined. There are, to be sure, unknown variable factors involved. For example, most of the examinations were probably of more favorable species of fishes. Mesopelagic fishes with less contact with molluscs will probably be found to harbor few Digenea. However, the figures do indicate the great prevalence and diversity of digenetic trematodes in marine fishes (mostly in tropical seas). If these fish actually harbor only 50 per cent of the reported number of Digenea, one might speculate that there are 10,000 species of Digenea in fishes. To date not more than 15 per cent of that number have been described. The Monogenea are no more completely known and are perhaps almost as numerous. Thus there is a great deal of taxonomic work to be done. Trematodes have enjoyed a long and successful coexistence with fishes and molluscs, and the evolution of all three groups has progressed together. Trematodes now unknown will surely supply information needed for understanding the systematics of the entire group.

The neglect of parasites as one of the large groups of marine life is so longstanding and so great that we should no longer be satisfied with a few parasitologists accompanying other expeditions to distant places. The need is for parasitological expeditions on which ichthyologists and malacologists accompany the parasitologists, rather than vice versa.

It might be noted incidentally that the search for and discovery of these parasites often entails labor and hardships never mentioned in the final publications. The work must be done while the host is freshly dead, wherever and whenever this may be. Such discomforts as seasickness and exposure to the elements must be disregarded. In 1922 Johnston and Tiegs published a monograph on some monogenetic trematodes of Australian freshwater fishes. It contains no mention of hardships involved in collecting these parasites, but the hosts could have been obtained only by travels into dangerous regions of the "outback." The type localities are still practically inaccessible. To reach them, one would need to charter a plane and endure hazards of desert life in remote, uninhabited, and inhospitable country.

A prerequisite to the collecting of any parasite is obtaining the host animal. Sometimes more time and effort are spent acquiring the hosts than in collecting the parasites, especially since it is highly desirable that hosts be examined soon after death before the parasites have started to disintegrate. The hosts themselves may well—in fact preferably—be collected and identified by vertebrate specialists, but the parasites should be collected by experienced parasitologists. Poorly preserved trematodes are almost useless.

A host animal may sometimes be cooled by ice or refrigeration for a period of 24–48 hours and still yield useful specimens of trematodes. Our experiences with trematodes from *fully* frozen fishes have been unsatisfactory. Worms are usually contracted, collapsed, and distorted, and are often more or less macerated. In fact, none of the various shortcuts proposed for collecting and preserving trematodes has proved desirable in general, although some may succeed with certain species. The preferred method is to kill individual specimens on a slide under a coverglass to which is applied enough pressure to flatten the specimen to normal extension, or to what appears to be normal. There are no easy ways, so far as I know, to collect large numbers of digenetic trematodes effectively and quickly, or even at one's convenience. On the other hand, the handling of well preserved specimens is relatively simple. They stain well with many reagents and beautiful preparations are easily possible.

Parasitologists have more chances to make mistakes than most other biologists, and they probably do so. For example, they may misidentify the host as well as the parasite. There are records of parasites from countries where the hosts involved never occur naturally. These records become listed in reliable references and may be very misleading. The monogenetic trematode genus *Diplozoon* Nordmann, 1832, is common in Europe and Asia; it occurs in Africa and probably in South America, but not in North America. In 1951 Reichenbach-Klinke described *Diplozoon barbi* taken from *Barbus semifasciolatus* and *Puntius tetrazona;* the hosts were obtained from the aquarium of a fish dealer in New Jersey and were examined in Germany. Both of these fishes are strictly Asiatic in distribution. They were only in transit through North America, a fact made clear in the published description. The *Index-Catalogue of Medical and Veterinary Zoology* (1964) and Yamaguti (1958) state the site of occurrence as North America, while the *Zoological Record* (1951) records Germany. Both are misleading.

A similar example is found in Travassos (1934) where an amphistome, *Helostomatis helostomatis* (MacCallum, 1905) was recorded as a South American parasite of *Helostoma. Helostoma* (Kuhl) is a fish of tropical Asia introduced as an aquarium fish in many parts of the world.

By the same token, records of parasites from animals in zoos are suspect. These parasites are likely to have been introduced with the zoo animal, but animals in captivity may become abnormally infected with parasites from animals about them.

Records of collections are often incomplete. They should include the number of hosts examined, number infected, intensity of infection, scientific and common name of host, family of host, exact location of host, and exact location of the parasite.

The disposition and availability of type specimens is another problem. Many authors, particularly in Asian countries, do not deposit holotypes in museums and some do not respond to requests for loan. Firm international rules or understandings should be imposed here.

There are, at present, a few troublesome problems in terminology which should be resolved. For example, the so-called sucker ratio of Digenea states the size of the oral sucker (taken as 1) compared with that of the acetabulum. Most authors apparently measure cross-diameters, but some average the length and the width of each sucker and thus compare general size rather than cross-diameters. Since the suckers are often not exactly circular and may be considerably greater in one dimension, the two procedures may give different results. In comparing descriptions of species one needs to know how the sucker ratios were derived, but this information is usually not given. Thus a valuable distinction between species is clouded with confusion.

To distinguish between "a tubular excretory vesicle" and a Y-shaped vesicle is not realistic. All excretory vesicles are tubular, whether I-shaped, Y-shaped, or V-shaped.

More precise definitions are needed for such terms as forebody, genital atrium, *pars prostatica,* and hermaphroditic sac, all of which have been used in more than one sense.

Among specialists on the Monogenea there are unfortunate differences in terms applied to hooks of the haptor and to certain terminal genital structures.

Literature

The literature on systematics of trematodes is extensive and widely scattered—so much so as to be a discouraging handicap to parasitologists without access to large or specialized libraries. The same situation exists for helminths in general. In 1941 Miriam Rothschild found that in a twelve month period, descriptions of 776 new species of helminths were scattered in 112 different journals. The situation today is probably worse. It is difficult to arrange publication of any lengthy monograph or review. Studies on collections of parasites must be published in parts, often in different journals. Costs of publication have led most journals to impose substantial page charges. Taxonomic papers are necessarily rather technical and of immediate interest chiefly to specialists, but they are important if we are to understand the whole picture of the living world and its interrelationships. Perhaps the best solution at present would be increased publication of longer papers by museums and educational institutions.

Various remedies for the literature problem in parasitology have been proposed and attempted. By far the most useful and comprehensive of these efforts has come from the Animal Disease and Parasite Division of the Agricultural Research Service at Beltsville, Maryland. The *Index-Catalogue of Medical and Veterinary Zoology* initiated by Stiles and Hassall (1902–1912) is being brought up to date and kept up to date by regular supplements. Thus references to literature in parasitology by author are available at reasonable cost. A modernization of the *Subject-Index* (by genera) started in 1963 with the Trematoda. Names of trematode genera, including a list of species in each genus, have now been published. Supplements to both the *Author-Index* and the *Subject-Index* are being issued annually. Such ambitious undertakings are of tremendous value.

Helminthological Abstracts, published (1932–) by the Commonwealth Bureau of Helminthology, St. Albans, England, is extremely useful in providing abstracts of papers including those in foreign languages. The *Zoological Record,* Vermes Section, has long been another valuable source.

All these aids are useful only if one knows the name of the author or the name of the parasite, for they are bibliographic and may not help much in the identification of a specimen. If an investigator decides upon the wrong family his specimen will easily seem to be a new genus. Thus, an important question is how can the literature help in the identification of specimens? The best aids for identification of trematodes in general have been: (1) *The Trematoda* by Dawes (1946); (2) the "Skrjabin Volumes" (*Trematodes of Animals and Man*) of which twenty-two volumes have now been published (1947–1966) containing a total of more than 15,000 pages; and (3) Yamaguti's *Systema Helminthum,* Vol. 1 (two parts): *Digenetic Trematodes* (1958). Three of the Skrjabin volumes have been translated from the Russian into English. Their value is obvious and the entire trematode series should be translated as soon as possible.

Early obsolescence is one serious difficulty with all such compilations. They begin to be out of date immediately after publication. Skrjabin tries to overcome this difficulty by repeating and updating families in later volumes. The result is trouble in assembling a family: one family would be scattered through several volumes, and increasingly so. The recent volumes are largely supplements in which families are without orderly arrangement. New editions and supplements to Yamaguti's *Systema Helminthum* would have some of the same failings. One would soon need compilations of the compilations.

In addition, compilations tend to be uncritical. As a result, nearly everything is accepted as originally published. Errors thus become more firmly established in widely used volumes. Even small errors, particularly in a key, may be very misleading.

The best solution to the problem of correct identification of trematodes is, I believe, in addition to the aids mentioned above, a loose-leaf notebook system in which the various taxa are described on separate pages. Such a system is not only flexible and subject to continual updating, but it is economical and subject to annotations, corrections, or replacement by the page. Several attempts have been made in this direction, the most successful of which, it seems to me, is that of the American Society of Ichthyologists and Herpetologists with the aid of the National Science Foundation. Printed sheets dealing with genera and species of Amphibia and reptiles are available at a cost of ten cents each. For example, a sheet on the salamander genus *Hydromantes* Gistal includes a Definition, Key to Species, Distribution, Comments, and a List of References. To deal with trematodes in a similar way would be a much larger project, but need not all be done at once. Such a system would have, of course, some of the weaknesses of any compilation, but it would have flexibility and could be easily revised.

Such a compendium of trematodes is so far in the future that some more immediate action is needed. A basic problem in the systematics of trematodes is the possible ease and low cost of collecting these parasites contrasted with the unavailability and high cost of adequate literature. As a result, very interesting material is collected, studied, and published without due regard to obscure or to recent literature. It seems almost hopeless to attempt to build up adequate libraries. Much of the literature is out of print and unavailable at any cost. Instead, individual investigators should be able to visit established libraries. In other words, parasitologists in all parts of the world should be encouraged to collect trematodes, but the final studies should be made in some center provided with adequate libraries. For trematodes there are not many such places—I should judge about eight in Europe, eight in the United States, one in Mexico, one or two in Japan, one or two in Australia.

In our experience we have found that actual specimens are often even more helpful than published descriptions, which may be incomplete or erroneous. Museums only partially meet this need; they offer holotypes and perhaps a few paratypes, but not a collection of species in general. The ideal center for taxonomic studies would be one with a good basic library, a loose-leaf set of descriptions, and a large, catalogued collection of specimens.

Taxonomy

The digenetic trematodes are a very diverse group of animals. Yamaguti (1958) recognized some 102 families and more than 1,000 genera. Generalizations are difficult in such a diverse group.

Even at the species level certain characters, such as extent of vitellaria, may be variable in some species while quite constant in others. The extent of variation in a species is sometimes not fully realized because of the small number of specimens available for study. As a rule only a few specimens occur in a host. It seems best to describe these as intelligently as possible pending later collections of additional specimens. Descriptions based on specimens of poor quality are less justified.

Host-induced variations may be a complicating factor, especially if the intensity of infection is abnormally high or if the host is not a normal one. Wolfgang (1955) found great variation in the oral spines of *Stephanostomum baccatum* (Nicoll, 1907) but based his conclusions on 400 specimens from a single host. Most species of this genus occur as from one to five specimens in a host, and in these cases oral spination seems rather constant.

Experimental study of variations induced by living in unnatural hosts has been made in only a few cases. Watertor (1967) reports that *Telorchis bonnerensis* Waitz, 1960, a parasite of salamanders, showed variation in body size, extent of vitellaria, and in egg size in certain species of salamanders and in turtles. Grabda-Kazubska (1967) found that the trematode, *Opisthioglyphe ranae* (Frolich, 1791), varied considerably according to the species of its final host. The most appropriate host was *Rana esculenta*. Both experimental and natural infections in *R. temporaria* and in *R. terrestris* consisted of specimens with shorter and wider bodies, testes of different shape, and different vitellaria. Infections in a snake, *Natrix natrix*, showed still other variations. As a result two species of *Opisthioglyphe* were reduced to synonymy with *O. ranae*. It seems doubtful if such host-induced variations are common in nature, at least in the case of older, well established species.

The higher taxa of trematodes are not so clearly determined as we once thought. The Digenea and Monogenea are, of course, clearly distinct. The trend, especially in Europe, is to consider each of these groups a separate class.

The Aspidogastrea, once considered as intermediate between Monogenea and Digenea, are definitely allied to the Digenea, even though they do not reproduce as larvae. The morphology of the adult, the total absence of haptoral hooks, and the use of molluscan hosts, all suggest the Digenea. Stunkard (1962) notes such affinities and, retaining names first proposed by Burmeister in 1856, places the orders Aspidobothrea and Digenea under the subclass Malacobothridia which, with the subclass Pectobothridia (= Monogenea), would constitute the class Trematoda.

If our classification is to be primarily a "natural" one based on phylogeny, some changes at the class level do seem justified; for example, class rank for the Monogenea, now commonly Class Monogenoidea. This action, of course, raises problems of terminology since both groups have been known as trematodes for so many years. Llewellyn (1965) suggests that the name "trematode" should not be retained. It will surely remain as a common name and could be used as a class name in place of Digenoidea, thus exiling the "monogenes" as non-trematodes.

Still assuming that a classification should reflect phylogenies, subdivisions of the Phylum Platyhelminthes are suggested on page 102.

This classification is largely based on relationships proposed by Llewellyn (1965). The *Gyrocotyle* group of "cestodaria" has true lycophore larvae and seems clearly related to the monogeneans. Dubinina (1960) showed that the larvae of *Amphilina* have four hooks of different origin than the other six.

The following classification considers (1) that the precocious reproduction of *Gyrodactylus* is essentially different from that of the larvae of Digenea; (2) that some Digenoidea (that is, aspirogastrids) may not reproduce as larvae;

and (3) that the Mesozoa are related to the Digenea and are not primitive Metazoa.

Phylum Platyhelminthes

Class Turbellaria
Class Nemertinea
Class Monogenoidea
Class Gyrocotyloidea
Class Digenoidea (or Trematoda)
  Subclass Aspidogastrea
  Subclass Digenea
Class Mesozoa
Class Çestoidea
  Subclass Amphilinidea
  Subclass Cestoda

Regarding the Digenea, the trend at the present time is to accept the classification proposed by LaRue (1957). This classification has been modified as new discoveries are made and will surely be further revised and extended. It perhaps overstresses the value of characters of larval stages.

Most of the changes in classification at the level of the family and above have come as a result of discoveries of life cycles. In a few cases the morphology of the adult was found to be misleading or to be less significant than characteristics of larval stages. Thus, the Gasterostomata were found to be less isolated than once thought and actually to be related to blood flukes and strigeids. The once isolated Bivesiculidae have been found (LeZotte, 1954) to be related to the Azygiidae, something that would never be suspected from the morphology of the adults. However, convergence, contradictions, and unexpected conditions have been found to occur among cercariae, and even the excretory system and flame cell patterns are not always reliable. It seems evident now that any or all stages in the life cycle might be phylogenetically important and our gradually improving classification must consider them all.

Any valid classification of trematodes must depend on evaluation of as many characters as possible. Still reliable and most important are those derived from comparative morphology, life cycles, host relations, and geographical distribution. When these factors are correctly known they usually fit together to form a consistent whole. Additional characters are now being supplied by electron microscopy, biochemical studies, and computer analyses. It is important to recognize that these new approaches give additional characters to be considered and that they are not replacements of the older classical methods. In fact, the new approaches must start with the product of a great amount of taxonomic work. So far as trematodes are concerned, much basic taxonomy remains to be performed and many errors remain to be corrected.

In summary, the difficulties we now face in the systematics of trematodes may be traced to (1) incomplete knowledge; (2) widely scattered literature; (3) erroneous and incomplete descriptions. Perhaps many of these difficulties go back to publication of poor work by inexperienced investigators. Yet papers by a beginner are often of excellent quality, and early publication of interesting material should always be encouraged. At the same time, adequate training in taxonomy, now neglected, should be promoted.

The final classification to be attained in the distant future will be the result of the dedicated and unselfish labor of generations of systematists. It will form a harmonious whole not only *per se* but with significant contributions to other aspects of biology such as zoogeography, paths of dispersal, places of origin, phylogeny of hosts, and ecological conditions both of the present and of the ancient past.

REFERENCES

Dawes, B. 1946. *The Trematoda.* Cambridge University Press, Cambridge.

Dubinina, M. N. 1960. The morphology of Amphilinidae (Cestodaria) in relation to their position in the system of flatworms. *Dokl. Akad. Nauk S.S.S.R.* 135: 501–504; 943–945.

Grabda-Kazubska, B. 1967. Morphological variability of adult *Ophisthioglyphe ranae* (Frölich, 1791) (Trematoda, Plagiorchiidae). *Acta Parasitol. Polon.* 15: 15–34.

Johnston, T., and O. W. Tiegs. 1922. New gyrodactyloid trematodes from Australian fishes, together with a reclassification of the super-family Gyrodactyloidea. *Proc. Linn. Soc. N.S. Wales* 47: 83–131.

LaRue, G. R. 1957. The classification of digenetic Trematoda: a review and a new system. *Expl. Parasitol.* 6: 306–349.

LeZotte, L. A., Jr. 1954. Studies on marine digenetic trematodes of Puerto Rico: the family Bivesiculidae, its biology and affinities. *J. Parasitol.* 40: 148–162.

Llewellyn, J. 1965. The evolution of parasitic Platyhelminthes. In: Angela Taylor, Ed. *Evolution of Parasites,* pp. 47–78. Blackwell, Oxford.

Skrjabin, K., *et al.* 1944–1966. *Trematodes of animals and man.* 22 volumes. Akademiya Nauk S.S.S.R., Moscow [In Russian].

Stunkard, H. W. 1962. *Taeniocotyle* nom. nov. for *Macraspis* Olsson, 1869, preoccupied, and systematic position of the Aspidobothrea. *Biol. Bull.* 122: 137–148.

Travassos, L. 1934. Synopse dos Paramphistomoidea. *Mem. Inst. Oswaldo Cruz.* 29: 19–178.

Watertor, J. L. 1967. Intraspecific variation of adult *Telorchis bonnerensis* (Trematoda: Telorchiidae) in amphibian and reptilian hosts. *J. Parasitol.* 53: 962–968.

Wolfgang, R. W. 1955. Studies of the trematode *Stephanostomum baccatum* (Nicoll, 1907). IV. The variation of the adult morphology and the taxonomy of the genus. *Can. J. Zool.* 33: 129–142.

Yamaguti, S. 1958. *Systema helminthum.* Vol. I. *The Digenetic Trematodes of Vertebrates.* Interscience, New York.

Discussion

Question: Your idea of a loose-leaf notebook is very interesting. Have you started such a project?

Dr. Manter: I started one about forty years ago. It is far from complete. But I do have about 140 notebooks full of descriptions and illustrations of digenetic trematodes. Although they took a great deal of time to compile, they now save me many hours of time while searching the literature. We have been urging Dr. Yamaguti to publish his revision of *Digenetic Trematodes* in loose-leaf form. Its publication has not yet been arranged, but we are hoping it will appear on separate, notebook-sized sheets. The advantage is that if you are interested in a given family, you can buy just those pages. If you are interested in a certain genus, you can buy everything that's known about that genus. To compile such a system would be very laborious, of course. Such information published in a volume would soon be out of date; but if it is loose-leaf, the pages can be rearranged or discarded when superseded by new information.

Question: Do you have any plans for publishing your 140 notebooks?

Dr. Manter: No. It would involve considerable polishing. However, if it meant publishing one genus or one family at a time, it might be possible.

Norman D. Levine, Director of the Center for Human Ecology and Professor of Veterinary Parasitology and Zoology at the University of Illinois, has held office in almost every American professional society dealing with parasites. He is President of the American Microscopical Society, President of the American Society of Professional Biologists, and Past President of both the Illinois State Academy of Sciences and the Society of Protozoologists. Currently he is Editor of Journal of Protozoology.

Norman D. Levine / Problems in Systematics of Parasitic Protozoa

Anton van Leeuwenhoek was the first person to describe a parasitic protozoon. He recorded *Eimeria stiedae* oocysts in a rabbit in 1674. Linneaus did not include this or any other parasitic protozoon in the 1758 edition of his *Systema Naturae*. Two hundred years later, however, the names of 6,775 such organisms had been listed in the *Zoological Record* (Levine, 1962). These included 212 sarcodines, 1,300 flagellates, 3,513 "Sporozoa," 1,700 ciliates, and 50 incertae sedis. In addition, 20,182 fossil species (the great majority foraminifera) and 17,293 free-living species were noted. We are concerned here only with the parasitic protozoa.

How many species of parasitic protozoa are there? During 1949–1958, an average of 115 new species was being named a year (Levine, 1962). On this basis one could estimate that by 1968 about 7,900 species had been named. But this is a small proportion of the actual number. In 1961 I estimated (Levine, 1962) that there might be 3,500 different species of the single genus *Eimeria* in mammals and perhaps 34,000 in chordates. If we assume (whether safely or not) that every species of animal has at least one species of parasitic protozoon, and if we accept the estimate of Muller and Campbell (1954) that there are about 898,450 species of living animals (exclusive of the Protozoa), then we can estimate that there must be almost 900,000 species of parasitic protozoa. In other words, names have been given only to about 1 per cent of the total. Whether this estimate is right or wrong, it is clear that we have quite a way to go in naming parasitic protozoa.

Supported by National Science Foundation Grant GB-5667X.

Classification

Now, how about the classification of parasitic protozoa? We have seen that they belong to all the major groups of protozoa.

Protozoa were not included in the 8th (1859–1860) edition of the *Encyclopedia Brittanica.* In Volume XIX (1885) of the 9th edition, however, they were accorded a long (36 page) article by E. Ray Lankester. He did not attempt to say how many species there were, but he divided the Protozoa into two "Grades," the Gymnomyxa and the Corticata. Under the grade Gymnomyxa, he included the classes Proteomyxa, Mycetozoa, Lobosa, Labyrinthulidea, Heliozoa, Reticularia (now known as Foraminiferida), and Radiolaria. Under the grade Corticata he included the classes Sporozoa, Flagellata, Dinoflagellata, Rhynchoflagellata (for *Noctiluca,* now known to be a dinoflagellate), Ciliata, and Acinetaria (for Suctoria and related forms, now known to be ciliates).

It is hardly worthwhile to give Lankester's classification in detail, since it is too unfamiliar and too incomplete. Inspection of a single class, the Sporozoa, will give its flavor. In this class he recognized four subclasses—Gregarinidea, Coccidiidea, Myxosporidia, and Sarcocystidia. He attributed all of these subclasses to Bütschli (1883). In the Gregarinidea he accepted two orders, Haplocyta Lankester and Septata Lankester (both new, I gather). The order Haplocyta contained a single genus, *Monocystis,* and the order Septata contained two, *Gregarina* and *Hoplorhynchus.* We now accept about 150 genera of gregarines, although I suspect that some of them are not valid. The subclass Coccidiidea contained three orders. In the order Monosporea was the single genus *Eimeria,* and in the order Oligosporea the single genus *Coccidium.* We know now that these "genera" are different stages in the same life cycle, and *Coccidium* has fallen as a synonym of *Eimeria* and other genera. Lankester's third order, the Polysporea, contained the single genus *Klossia,* with three species. The definitions and names of the orders of Coccidiidea have changed considerably, and so have the number of genera.

Lankester did not even include *Plasmodium,* the cause of malaria, for it was not yet known. He said, "No disease is known at present as due to Sporozoa," a statement which clearly reveals the state of our knowledge of the Protozoa in 1885. Indeed, so far as I am aware, the only pathogenic protozoon known at the time was *Trypanosoma evansi,* which was found by Evans in India in 1881, where it was causing a disease in elephants known as surra. The protozoan nature of the causative agents of African and American trypanosomosis, leishmaniasis, malaria, coccidiosis, babesiosis, theileriosis and, of course, gregarine infections of insects, was as yet unknown, as were the parasites themselves.

Hall (1953) reviewed the classification of protozoa, and I shall not repeat the details here. Perhaps the first fairly modern classification of the phylum was that of Kent (1880–1882). He recognized the classes Rhizopoda (containing the gregarines, now known to be Sporozoa), Flagellata (containing the sponges), Ciliata, and Tentaculifera (Suctoria and Actinaria). The class Sporozoa was established by Leuckart (1879) ; he included in it only the gregarines and coccidia, but of course the malaria parasites were not known at the time. I have already discussed Lankester's (1885) classification. Bütschli (1880–1889) accepted the classes Sarkodina, Sporozoa, Mastigophora, and Infusoria (for the ciliates and Suctoria). Other individuals made different contributions. Perhaps the most significant was that of Doflein (1901), who separated the Protozoa into two subphyla, Plasmodroma (containing the classes Mastigophora, Sarcodina, and Sporozoa) and Ciliophora (containing the classes Ciliata and Suctoria) on the basis of the type of nucleus (vesicular versus macronuclei and micronuclei) and absence or presence of cilia. This basic scheme has been used in most textbooks. It was more or less crystallized by a group of American protozoologists, who prepared a classification of the Protozoa for the American Association for the Advancement of Science in 1936. Later research, however, revealed that this classification was far from satisfactory, and the need for further overhauling was recognized.

The best present classification of the Protozoa is that proposed by a committee of the Society of Protozoologists in 1964 (Honigberg *et al.,* 1964). It was discussed at some length at the Second International Conference on Protozoology in London in 1965 (see Levine, 1966). Other recent classifications which differ more or less from this one include those of Grassé (1952–1953), Piekarski (1954), Grell (1956), Kheisin and Polyansky (1963), and Raabe (1964). The Society's classification divides the phylum into four subphyla. The Sarcomastigophora have flagella, pseudopods, or both. They have a single type of nucleus except in the developmental stages of certain Foraminiferida. If sex is present, fertilization is by syngamy (that is, gametes are produced which fuse). Typically, there are no spores. This subphylum is divided into the superclass Mastigophora, with flagella; the superclass Opalinata, with cilia-like organelles; and the superclass Sarcodina, typically with pseudopods.

The second subphylum is the Sporozoa. These organisms typically have spores, but the spores contain no polar filaments. There is a single type of nucleus. There are neither cilia nor flagella, except that the microgametes are flagellated in some groups. Sex is ordinarily present, and fertilization is by syngamy. All the Sporozoa are parasitic. This subphylum contains the following classes: Telosporea, which is typical of the subphylum and which contains the subclasses Gregarinia and Coccidia; Toxoplasmea, in which there are no spores and no sex; and Haplosporea, in which there are spores but no sex.

The third subphylum is the Cnidospora, in which the spores have one or more polar filaments and one or more sporoplasms, and all of whose members are parasitic. This subphylum contains the class Myxosporidea, with spores of multicellular origin; and the class Microsporidea, with spores of unicellular origin.

The fourth subphylum is the Ciliophora. Its members have cilia, generally two types of nucleus, and sexuality involving conjugation. It has a simple class, Ciliatea, which includes the following subclasses: Holotrichia, with somatic ciliature generally simple and uniform; Peritrichia, with somatic cilia essentially absent in the mature form (but oral ciliature conspicuous); Suctoria, with no somatic or oral cilia but with tentacles in the adult; and Spirotrichia, with somatic cilia generally sparse.

I shall say nothing about the ciliates here. The great majority are free living and the systematics of the parasitic members is quite well established.

Opalinids

The first problem I should like to mention is that of the opalinids. These are intestinal parasites of Amphibia. They have cilia (or cilium-like) organelles in oblique rows over their entire body surface. They have no mouth. They have from two to many nuclei of one type and divide by binary fission. So far as is known, their life cycles involve syngamy with flagellated gametes. This group has conventionally been placed under the ciliates because its members have cilia, but in all other respects it belongs elsewhere. And the electron microscope has revealed that there is no basic difference between cilia and flagella, so that it is not appropriate to divide the Protozoa into two groups on the basis of whether they have one or the other. I myself think that the type of nucleus is far more appropriate as a determining character, and I think that a basic division could well be into the forms with a single type of nucleus (that is, a vesicular nucleus) and the forms with two types of nucleus (macronucleus and micronucleus). This would leave the so-called ciliates together as before, but would remove the opalinids. As already mentioned, the Society of Protozoologists' classification recognized the uniqueness of this group, and placed it as a superclass under the subphylum Sarcomastigophora. However, this was a compromise action, and is subject to modification. One can think of this group as homokaryotic ciliates (that is, with a single type of nucleus), or as flagellates whose flagella have gotten smaller and turned into cilia. I prefer the latter, but some people do not. The Society of Protozoologists committee compromised by considering them to be intermediate between flagellates and ciliates, and then established a separate superclass,

Opalinata, for them which the committee attached to the subphylum Sarcomastigophora. This is as good a temporary solution as any.

Piroplasmids

The second problem has to do with the piroplasmids. They have been traditionally considered as some sort of appendage to the Sporozoa, although their affinities were far from clear. However, some modern taxonomists thought that they really belonged with the Amoebae, and the Society of Protozoologists' classification placed them in a separate class, Piroplasmea, under the superclass Sarcodina. This was reasonable at the time, but recent electron microscope work has revealed that they have a polar ring, micronemes, rhoptries, and other organelles similar to those of the coccidia, malaria parasites, *Toxoplasma,* and *Sarcocystis* (Simpson, Kirkham, and Kling, 1967; Friedhoff and Scholtyseck, 1968). This means that they belong in the same group with these forms; that is, to the so-called Sporozoa.

The reason that I call them the "so-called Sporozoa" is that the name implies that members of this class have spores. This is true of the coccidia and the gregarines, which were the original members of the group, but it is not true of the piroplasmids, toxoplasmids, and sarcocystids, and it is also not true of the malaria parasites and their relatives. As Gordon Ball (1960) said, "probably the only character they all possess is a parasitic mode of life, hardly a distinguishing trait separating them from other kinds of the Protozoa." What we need, obviously, is a new name for the group, a name which recognizes their common characteristics and sets them off from all other Protozoa. I think that such a name should be based on some structure revealed by the electron microscope, but I hesitate to suggest one until we find out by such studies what structures the gregarines have. Perhaps we shall have some information on this point in a year or so.

To return to the piroplasmids, one of the problems that plagued protozoologists for years (and that is still perhaps unresolved) has to do with sex. We know that the life cycles of malaria parasites and their relatives have both sexual and asexual phases which alternate with each other. The same is true of the coccidia. Since the piroplasmids are similar to the malaria parasites in many ways, it was natural to assume that they, too, have sexual and asexual phases in their life cycles. And some investigators saw what they expected to see when they looked into the microscope—they saw forms which could be interpreted as sexual stages. I discussed their work in some detail in an earlier publication (Levine, 1961), and I have also described more recent work which contradicts these earlier interpretations; therefore I will not repeat what I said here. Suffice it to say that there is no reliable evidence that sex occurs in the piroplasmids.

Sporozoa

The third problem has to do with the so-called Sporozoa whose spores have a polar filament. This feature differentiates them from all other members of the group, and the electron microscope has revealed that they do not have a polar ring, conoid, rhoptries, micronemes, and other organelles which the Sporozoa possess. Obviously they are different. The Society of Protozoologists removed them from the Sporozoa and set up another subphylum, the Cnidospora, for them. But this action, although superficially proper, did not solve the whole problem. Let me explain. There are two classes within the Cnidospora. In the Myxosporea the spore is of multicellular origin; it contains several nuclei and usually contains more than one polar filament. The sporoplasm (the small amoeboid organism contained in the spore) or sporoplasms emerge through an opening at the base of the extruded polar filament. In the Microsporea the spore is of unicellular origin. There is a single polar filament, a single sporoplasm, and the sporoplasm emerges through the polar filament and not at its base. Do these two groups belong together? Some authors such as Ulrich (1950) have removed the Myxosporea from the Protozoa entirely, and have placed them in a group intermediate between the Protozoa and Metazoa. Very few protozoologists are willing to go this far, and I myself am unwilling to do so, at least at present. However, the problem should be mentioned. I have a feeling that these two groups may well have originated separately and have been placed together primarily as a matter of convenience.

Haplosporea

The fourth problem has to do with the Haplosporea. This is a group of parasites of molluscs and other invertebrates which have spores but no polar filaments, polar rings, conoids, micronemes, rhoptries, flagella, or cilia. They reproduce by schizogony; there is a difference of opinion whether sex is present. The Society of Protozoologists (Honigberg *et al.,* 1964) placed this group as a class in the Sporozoa, but as a matter of expediency rather than of conviction. Indeed Weiser (1964) thought that there was a close relationship between the Cnidospora and the Haplosporea. Sprague (1966) returned to the old separation of Delage and Hérouard (1896) and rearranged the "Sporozoa" and "Cnidospora" in two subphyla, Rhabdogena and Amoebogena, based on the shape of the germinal elements or sporoplasms in their spores; the subphylum contains both the forms with polar filaments (that is, the Cnidospora) and the Haplosporida. This grouping has the advantage of removing the Haplosporida from the Sporozoa, leaving the latter group more uniform, but it provides no logical place for the piroplasmids. These have a polar ring, rhoptries, and so on, thus obviously belonging with the Sporozoa, but their germinal elements are not rod shaped or crescent shaped. Hence calling the "Sporozoa" Rhabdogena will not do.

Weiser (1966) returned to Schaudinn's (1900) Neosporidia, including in it the Microsporidia and Haplosporidia (Schaudinn had included the Microsporidia, Myxosporidia, Actinomyxidia, Sarcosporidia, and Haplosporidia). However, since Weiser was concerned in his book only with parasites of insects, he included no other groups. I think that Weiser's approach is probably the most logical at this time, but of course it should be expanded to include the Myxosporea. The modification that I am therefore suggesting is to name the subphylum "Neospora Schaudinn, 1900," and to include under it the three classes Myxosporasida Bütschli, 1881, Microsporasida Corliss and Levine, 1963, and Haplosporasida Caullery, 1953. Perhaps eventually we should change this by removing the Myxosporasida to the Mesozoa.

Gregarines

The fifth problem, which is related in a way to the preceding ones, has to do with the gregarines. These are unquestionably Sporozoa, but so far no one has seen a polar ring, conoid, micronemes, rhoptries, or other characters in them. But no one has looked at their sporozoites with the electron microscope. So we must wait a while for the answer to this question.

The classification of the gregarines and the names of the various stages in their life cycles require mention. The gregarines are closely related to the coccidia, and have a similar life cycle, involving sexual reproduction. But different terms have been used for what are essentially the same stages. This was because different persons worked on the two groups, and each used his own terminology. We must therefore develop a set of standard names for sporozoan stages and structures.

We must also work out a rational classification of the gregarines. Something more than 150 genera have been named, but I doubt very much if all of them are valid. Most have been named on the basis of the characteristics of a single stage in the life cycle, and the life cycles of most of them have not even been worked out. There are not many people working on gregarines today, and many of the species have not been seen since they were first named, often more than 50 years ago. We need a thorough revision of the group, but this will require a great deal of work.

Position of Genera

A sixth problem has to do with the position of certain genera. I suppose that there are more genera of so-called Protozoa than of other phyla whose taxonomic position is uncertain. We are slowly decreasing their numbers, but we have quite a distance to go. We know now that *Bartonella* is a bacterium, that *Anaplasma, Eperythrozoon,* and *Haemobartonella* are rickettsiae, and that *Toxoplasma, Babesia,* and *Theileria* are true protozoa. *Pirhemocyton* is recognized to be a virus inclusion body, and *Toddia* has recently been found to be the same (Marquardt and Yaeger, 1967). Another enigmatic organism, *Pneumocystis,* has also yielded to placement. This organism, which causes an alveolar pneumonitis in rats and men, superficially resembles *Toxoplasma,* but recently Vavra, Kucera, and Levine (1968) found it to be a yeast. But there are other genera whose taxonomic position remains to be determined.

One Phylum, or Several?

A seventh problem, and a basic one, has to do with the question of whether the Protozoa are a single phylum or several. This question stems back to the last century and has many ramifications. Are the Protozoa cellular or acellular? I shall not go into this, but reference may be made to Corliss (1957). Should we speak of Protozoa or Protista? Haeckel (1866) introduced the latter term, and there is justice in the idea that protozoa are neither plants nor animals but something in between. Nevertheless, this view has not been well accepted—except philosophically—and the phytoflagellates and other plant-like forms continue to be accepted as plants by the botanists and as animals by the zoologists. Reference should be made to Honigberg (1967) for details and the literature.

Should we consider the Protozoa (or Protista) a separate kingdom composed of several phyla (Kozloff, 1960), or should we consider them a single phylum? I think that this is immaterial at the moment. The basic question which all these other questions imply (but which most of their makers do not realize) is a semantic one, a verbal issue. It has to do with the origin of the protozoa and of their various groups. Some workers seem to have the idea that if we say that there are several phyla of protozoa, then we do not have to worry about their origin—or at least not as much as if we say that they compose a single phylum. This is not true. We must still concern ourselves with the origin of groups of organisms, whether we call the groups phyla, classes, orders, or anything else. And we have very little to go on for the Protozoa. There are no fossils except in a few groups. At present we think that the phytoflagellates are the most primitive and the parents of all the other groups. This is probably true, but not everyone agrees. And the origin of many groups is still a complete puzzle. Where did the Microsporea come from? Did the Myxosporea come from the same source? They both have polar capsules and polar filaments. How about the Haplosporea? Are they closer to the Microsporea than to the Sporozoa? Where do the opalinids fit? I could go on for quite a while, and other protozoan specialists could cite still other puzzles.

A physicist recently told one of my colleagues that he sometimes wished that he was working in biology, where all the problems had been settled. I only wish that his statement reflected knowledge rather than ignorance. Maybe in a few hundred years the situation will be better. But more than likely, by then the taxonomists and biologists will have other problems, equally difficult.

REFERENCES

Ball, G. H. 1960. Some considerations regarding the Sporozoa. *J. Protozool.* 7:1–6.

Bütschli, O. 1880–1889. *Protozoa.* Vol. I, pts. 1, 2, 3. In: Bronn's *Klassen und Ordnungen des Thierreichs.* C. F. Winte'sche Verlag, Leipzig.

Corliss, J. O. 1957. Concerning the "cellularity" or acellularity of the protozoa. *Science* 125:988–989.

Delage, Y., and E. Hérouard. 1896. *Traité de zoologie concrète.* I. Schleicher Frères, Paris.

Doflein, F. 1901. *Die Protozoen als Parasiten und Krankheitserreger.* Gustav Fischer, Jena.

Friedhoff, K. T., and E. Scholtyseck. 1968. Feinstruktur von *Babesia ovis* (Piroplasmidea) in *Rhipicephalus bursa* (Ixodoidea): Transformation sphäroider Formen zu Vermiculaformen. *Z. Parasitenk.* 30:347–359.

Grassé, P.-P., ed. 1952–1953. *Traité de zoologie. Anatomie, systématique, biologie.* Vol. I, fascicles 1, 2. *Protozoaires.* Masson, Paris.

Grell, K. G. 1956. *Protozoologie.* Julius Springer, Berlin.

Haeckel, E. 1866. *Generelle Morphologie der Organismen.* Berlin. 2 vols.

Hall, R. P. 1953. *Protozoology.* Prentice-Hall, New York.

Honigberg, B. M. 1967. Problems in classification of Protozoa. *Bull. Nat. Inst. Sci. India.* 34:13–19.

——— W. Balamuth, E. C. Bovee, J. O. Corliss, M. Gojdics, R. P. Hall, R. R. Kudo, N. D. Levine, A. R. Loeblich, Jr., J. Weiser and D. H. Wenrich. 1964. A revised classification of the phylum Protozoa. *J. Protozool.* 11:7–20.

Kent, W. S. 1880–1882. *A Manual of Infusaria.* [Privately published.] London.

Kheisin, E. M., and G. I. Polyansky. 1963. On the taxonomic system of Protozoa. *Acta Protozool.* 1:327–352.

Kozloff, E. N. 1960. Do protozoa represent a single phylum? Unpublished paper presented at a round-table discussion on "Problems in systematics and evolution of Protozoa" at the American Institute of Biological Sciences meeting, Stillwater, Oklahoma, August 1960.

Lankester, E. R. 1885. Protozoa. *Encyclopaedia Britannica,* 9th ed. Vol. XIX, pp. 830–866.

Leuckart, R. 1879. *Die Parasiten des Menschen.* 2nd ed. G. F. Winter, Leipzig.

Levine, N. D. 1961. Problems in the systematics of the "Sporozoa." *J. Protozool.* 8:442–451.

——— 1962. Protozoology today. *J. Protozool.* 9:1–6.

——— ed. 1966. Discussion of the classification of Protozoa at the Second International Conference on Protozoology, London, England. *J. Protozool.* 13:189–195.

Marquardt, W. C., and R. G. Yaeger. 1967. The structure and taxonomic status of *Toddia* from the cottonmouth snake *Agkistrodon piscivorus leucostoma. J. Protozool.* 14:726–731.

Muller, S. W., and A. Campbell. 1954. The relative number of living and fossil species of animals. *Syst. Zool.* 3:168–170.

Piekarski, G. 1954. *Lehrbuch der Parasitologie.* Julius Springer, Berlin.

Raabe, Z. 1964. Remarks on the principles and outline of the system of Protozoa. *Acta Protozool.* 2:1–18.

Schaudinn, F. 1900. Untersuchungen über den Generationswechsel bei Coccidien. *Zool. Jahrb., Abt. Anat.* 13:197–292.

Simpson, C. F., W. W. Kirkham and J. M. Kling. 1967. Comparative morphologic features of *Babesia caballi* and *Babesia equi. Am. J. Vet. Res.* 28:1693–1697.

Sprague, V. 1966. Suggested changes in "A revised classification of the phylum Protozoa," with particular reference to the position of the haplosporidans. *Syst. Zool.* 15:345–349.

Ulrich, W. 1950. Begriff und Einteilung der Protozoen. In: *Moderne Biologie. Festschrift 60. Geburt. Hans Nachtsheim,* pp. 241–250. Peters, Berlin.

Vavra, J., K. Kucera and N. D. Levine. 1968. An interpretation of the fine structure of *Pneumocystis carinii. J. Protozool.* 15 (Suppl.) :12–13.

Weiser, J. 1964. In *litt.* to Honigberg *et al.,* 1964.

——— 1966. *Nemoci hmyzu.* Academia, Prague.

Discussion

Question: Does your committee [of the Society of Protozoologists] include the slime molds in its classification?

Dr. Levine: We include them and the other plant-like organisms. But what do you do with some of these things? For example, you find an organism with chloroplasts, producing its own food by photosynthesis. It is obviously a plant, but if you treat it with a drug and destroy the chloroplasts it behaves like an animal. So we include the slime molds in our classification in the belief that they indicate the flagellate, photosynthetic ancestry of the protozoa.

Question: Do you think anything would be gained by subdividing the Protozoa into several phyla? Of course the Metazoa are already divided into several phyla, so why is this not true of single-celled organisms, and would any purpose be served by it?

Dr. Levine: I would have no objections to it but I do not know if anything would be gained. At least it would show that there are major differences between some groups. It is very difficult to get protozoologists to agree on any of these subjects. Americans think of the plant-like flagellates as the most primitive of the protozoa, but I heard a Russian authority, Oparin, say very seriously that we do not know what we are talking about—he said that the amoebae are the most primitive. And maybe he is right.

Question: In the absence of complete data on biology, such as life cycles, it seems to me that perhaps here is a situation where some of the numerical procedures would fit in well. You keep referring to electron microscope similarities and that sort of criterion. But has numerical taxonomy been seriously considered?

Dr. Levine: I do not think that this has been done for the Protozoa. I have been trying to do something of the sort for some of the Sporozoa but I have found that the structural information is so incomplete and so untrustworthy that I don't dare depend on this technique. Some of the species of gregarines, for example, have not been seen or described again since they were first seen more than fifty years ago. Every time someone studies the gregarines of earthworms, he finds many new species and even new genera, and does not find those species and genera that other workers have recorded. I hope that good anatomic studies will be made so that numerical procedures can be tried.

Gerald D. Schmidt / Conclusion

It is quite evident from the discussions that not all problems in systematics of parasites have yet been solved. Although each contributor discussed primarily the difficulties within his own field of research, it becomes apparent that most of these difficulties are common to all. For instance, it is agreed that in many cases the current taxonomic groupings in use are unnatural, particularly at the level of the higher categories. Yet attempts to reclassify major groupings based on newer information are often frustrated by the conservatism of systematists themselves.

One of the major problems that was pointed out is the paucity of young persons electing careers in parasite systematics. The reasons are evident: adequate training in systematics is not widely available, monetary funding is not easily obtainable for research of this nature, and employer's demands for zoologists with training heavily oriented toward biochemistry have discouraged graduate students from the areas of systematics.

It has been repeatedly noted that only a small fraction of the parasite species has been described, while new forms are being discovered at an increasing rate. This, together with the fact that the expert taxonomists are retiring faster than they are being replaced, indicates a crisis in the near future. Most practicing taxonomists today have a backlog of specimens which were sent to them by nontaxonomists for identification and which they despair of ever completing. The problem becomes even more acute when zoonoses due to undescribed parasites are discovered.

Ample evidence has been presented that in every group of parasites confusion abounds due to inadequate morphological studies. Particularly in the older literature, descriptions which were thought to be sufficiently detailed and accurate when written are confusing and inaccurate by today's standards. Much basic anatomic and taxonomic work must be performed to redefine these species, for they are for the most part the basis for our present scheme of classification.

At the same time it is recognized that new techniques must be developed to differentiate species and to describe them in such a way that they will not become mysteries for some future systematist. New parameters for differentiation must include all available information, be it mathematical, chemical, behavioral, ecological, or whatever. The Brave New Taxonomist, as Dr. Voge calls him, is faced with challenges comparable to any in science. We wish him courage and fortitude, for the future of biology depends in large part upon him.

Index

T